Elements of
STATISTICAL
THERMODYNAMICS

LEONARD K. NASH

Harvard University

Second edition

ADDISON-WESLEY PUBLISHING COMPANY
Reading, Massachusetts
Menlo Park, California · London · Don Mills, Ontario

This book is in the
ADDISON-WESLEY SERIES IN THE PRINCIPLES OF CHEMISTRY

Consulting Editor
Francis T. Bonner

ISBN 0-201-05229-6
ABCDEFGHIJ-AL-787654

Preface

Macroscopic thermodynamics is here reexamined from the perspective of atomic-molecular theory. The thermodynamic concepts of entropy and equilibrium are thus invested with new meaning and implication, and one comes to see how thermodynamic magnitudes (e.g., gaseous heat capacities and equilibrium constants) can be calculated from spectroscopic data.

Chapter 1 introduces and develops a statistical analysis of the concept of distribution—culminating in a very simple derivation of the Boltzmann distribution law and a demonstration of the relation of statistical and thermodynamic entropies. Chapters 2, 3, and 4 then treat in turn the formulation, the evaluation, and the application of partition functions.

Compared with the first edition, the present second edition offers in its opening chapter an analysis that is both much simpler and more decisive. This chapter also provides a brief but convincing demonstration of one crucial point merely announced in the first edition, namely: the enormous odds by which some distributions are favored over others. The remaining chapters reflect complete reorganization and extensive rewriting of the corresponding material in the first edition, and further incorporate several major additions. These include some illuminating interpretations of partition functions and statistical entropies, a brief but substantial development of the statistics of indistinguishable units, an exploration of the "dilute-gas condition" under which these statistics reduce to a limiting form equivalent to "corrected" Boltzmann statistics, and a derivation of the Maxwell-Boltzmann molecular-speed distribution law in three dimensions. Enlarged by 70%, the set of problems should both challenge and reward the reader.

This text has been designed to convey its message at either of two levels. First: if systematically supported by a series of exegetical lectures, the book may be used in strong introductory college chemistry courses. Years of experience in teaching such a course convince me that, wherever classical thermodynamics can be taught successfully, this much statistical thermodynamics can be taught at least as successfully—and with striking reflexive improvement in student understanding of classical thermodynamics. Second: the book lends itself to essentially independent study by more advanced students who seek a view of the subject that is less formal, and far less compressed, than that afforded by undergraduate physical-chemistry texts.

Irrespective of background, the reader will find need for just three small bodies of prior knowledge.

1. A rudimentary knowledge of the calculus, such as is acquired by my own freshmen in the first two-thirds of an introductory course in college mathematics.

2. An elementary understanding of macroscopic thermodynamics, on which we draw for a few simple relations like $dE = T\,dS - P\,dV$. Many current texts for the introductory college chemistry course offer a quite sufficient foundation in this area.

3. A slight acquaintance with the energy-quantization conditions that permit calculation of molecular parameters from spectroscopic measurements. The inclusion of some abbreviated didactic material lends the present book a minimal self-sufficiency in this department, but one may well prefer the more ample background afforded, for example, by pp. 1–80 of G. M. Barrow's *Structure of Molecules* (Benjamin, 1962).

The argument constructed on these modest foundations comprises essentially all the statistical mechanics that appears in even the most sophisticated of undergraduate physical-chemistry textbooks. I have neglected internal rotation, and I have not pursued the argument into the realm of kinetics—where it readily yields some powerful new insights. More significant than these easily remediable omissions is one notable limitation: applying to assemblies of effectively *independent* units, the results here obtained from analysis of microcanonical ensembles cannot at once be extended to assemblies of strongly interacting units. Given a sufficiently enlarged background, one may easily approach these important assemblies by way of Gibbsian analyses (of canonical and ground canonical ensembles) that fully display both the great power and the great beauty of statistical mechanics.

Among the many equations appearing herein, some are marked by letters to facilitate back-references in the immediately succeeding text, or in a problem. On the other hand, apart from facilitating such back-references, the numbers attached to thirty-odd equations signal a call for particular attention. As they proceed, readers would be well advised to compile their own lists of these numbered equations, each of which expresses an important idea and/or represents a useful computational tool.

Cambridge, Massachusetts   L.K.N.
August 1973.

Contents

Acknowledgments

I am obliged to the publishers cited below for permission to reproduce a number of figures taken from copyrighted works. As identified by the numbers assigned them in this book, the figures in question are:

Figure 7, taken from p. 55 of J. D. Fast's *Entropy* (Eindhoven, Holland: Philips Technical Library, 1962);

Figure 16, from p. 68 of G. W. Castellan's *Physical Chemistry*, 2nd ed. (Reading, Mass.: Addison-Wesley Publishing Co., 1971);

Figure 19, from an article by R. K. Fitzgerel and F. H. Verhoek, *Journal of Chemical Education* **37**, 547 (1960);

Figure 22, from p. 142 of Malcolm Dole's *Introduction to Statistical Thermodynamics* (Englewood Cliffs, N.J.: Prentice-Hall, 1954);

Figures 23 and 24, from pp. 203 and 257 respectively of *Statistical Thermodynamics* by J. F. Lee, F. W. Sears, and D. L. Turcotte (Reading, Mass.: Addison-Wesley Publishing Co., 1963); and

Figure 25, from p. 102 of *Statistical Thermodynamics* by R. H. Fowler and E. A. Guggenheim (Cambridge: University Press, 1939).

I am happy to acknowledge that my first 18 pages have been developed on a pattern suggested by reading of the late Ronald W. Gurney's ingenious introductory text. The arguments on pp. 49–53 and 61–62 owe some of their shape to constructive criticisms of the first edition forwarded to me by William C. Child, Jr., who has kindly supplied several more comments on a draft of this second edition. For a multitude of additional suggestions I am much indebted to Francis T. Bonner, Peter C. Jordan, and Lawrence C. Krisher. To Walter Kauzmann, the author of a splendid alternative to the present text, I am deeply grateful for his generous willingness to indicate many possible improvements. To my wife, Ava Byer Nash, I am beholden not merely for assistance with page-proofs but also for the life-support system that has sustained me throughout. All the foregoing have helped to make this book better than it could otherwise have been. Sole responsibility for any residual blemishes rests with the author, who will welcome any communications making known to him the error(s) of his ways.

The Statistical Viewpoint   1

In every change, however drastic it may appear, we surmise a "something" that remains constant. From the very beginning of the modern era, certain men (e.g., Descartes) have conceived that "something" in terms suggestive of what we would call energy. And energy—or, better, mass-energy—*is* surely conceived by us as a "something constant" enduring through all change. The energy concept thus gives quantitative expression to our firm conviction that "plus ça change, plus c'est la même chose." But we have too another conviction scarcely less intense: the conviction that the future will not repeat the past, that time unrolls unidirectionally, that the world is getting on. This second conviction finds quantitative expression in the concept of entropy (from Gr. *en*, in + *trope*, turning). By always increasing in the direction of spontaneous change, entropy indicates the "turn," or direction, taken by all such change.

From a union of the entropy and energy concepts, little more than a century ago, there was born a notably abstract science with innumerable concrete applications; a science of thermodynamics that combines magnificent generality with unfailing reliability to a degree unrivaled by any other science known to man. Yet, for all its immense power, thermodynamics is a science that fails to reward man's quest for understanding. Yielding impressively accurate predictions of *what* can happen, thermodynamics affords us little or no insight into the *why* of those happenings. Thus it permits us to calculate what is the position of equilibrium in the system N_2—H_2—NH_3, for example, but it fails entirely to tell us why that is the equilibrium condition for this specific system.

To be sure, given that certain thermodynamic parameters (the "free energies") are what they are, we readily see that a particular equilibrium condition is entailed. But we can find in thermodynamics no explanation of why the free energies are what they are. And in general, though thermodynamics teaches us to see important *relations* among the various macro-

1

scopic properties of a substance, so that many can be calculated from experimental measurements of a few, thermodynamics is powerless to produce from its own calculations numerical values for the few.

What is it about NH_3 that determines the magnitude of the free-energy characteristic of that compound? In principle this question should, we feel, be answerable. But we find scant prospect of any such answer in a classical thermodynamics which, focusing solely on the properties of matter in bulk, eschews all concern with the microcosmic constitution of matter. For consider that we can hope to *explain* the free energy of some substance only by showing how that particular free energy is entailed by the distinctive values of the atomic and/or molecular parameters of the substance. That is, given a (spectroscopic) determination of such parameters as the length, angle, and flexibility of the bonds in NH_3, we must be able to see that the free energy of NH_3 could not be other than it is. This will be possible only if we can bridge the gap between the microcosmic realm of atoms and molecules and the macroscopic realm of classical thermodynamics.

Statistical mechanics provides such a bridge, by teaching us how to conceive a thermodynamic *system* as an *assembly of units.* More specifically, it demonstrates that the thermodynamic parameters of the system are interpretable in terms of—and are indeed calculable from—the parameters descriptive of such constituent units as atoms and molecules. In a bounded system, the crucial characteristic of these microcosmic units is that their energies are "quantized." That is, where the energies accessible to a macroscopic system form a virtual continuum of possibilities, the energies open to any of its submicroscopic components are limited to a discontinuous set of alternatives associated with integral values of some "quantum number."

Perhaps the most familiar example of what is meant by quantization is presented by the Bohr interpretation of the hydrogen emission spectrum. This spectrum consists of a series of sharp "lines," characterized by particular wavelengths. Each of these lines is supposed to arise in the emission by the hydrogen atom of an energy packet of some particular size. Such an energy packet is emitted when the atom passes from a state of higher energy to one of lower energy. From a study of the sizes of the emitted energy packets, one infers that the atom can exist only in a certain well-defined set of quantum states. The energy (ϵ_H) associated with any of these permissible states is given by the equation:

$$\epsilon_H = -\frac{2\pi^2 me^4}{h^2} \cdot \frac{1}{n^2}.$$

Here h symbolizes Planck's universal constant, m and e respectively represent the mass and charge of the "orbital" electron in the hydrogen atom, and n

is a quantum number that can assume any *integral* value within the range 1 to ∞. The possible states of the hydrogen atom, each characterized by some integral value of the quantum number *n*, are thus linked with the discontinuous set of permissible energies given by the last equation—which expresses the *energy-quantization condition* for the hydrogen atom. Rather more complicated relations, involving additional quantum numbers, express analogous energy-quantization conditions applicable to other species of gaseous atoms.

Like atoms, molecules also can exist only in particular sets of states characterized by different electronic configurations, with which are associated correspondingly restricted series of permissible energy states. But, unlike atoms, molecules exhibit fully quantized modes of energy storage other than that represented by electronic excitation. For example, when in any given electronic state, a molecule may perform various vibrational motions. A study of molecular spectra indicates that, when the vibration can be approximated as a harmonic oscillation, the only permissible values of the vibrational energy (ϵ_v) are given by the equation

$$\epsilon_v = (v + \tfrac{1}{2})h\nu.$$

Here ν is a frequency characteristic of the particular vibration involved, and v is a quantum number that can assume any *integral* value within the range 0 to ∞. The possible vibrational states, each specified by some distinctive integral value of the vibrational quantum number v, are thus linked by the last equation with an evenly spaced set of quantized vibrational energies. The rotational motions of molecules, and the translational motions of both atoms and molecules, are similarly associated with sets of discrete quantum states—to which correspond similarly discontinuous series of permissible rotational and translational energies.

We seek then to view a macroscopic thermodynamic system as an assembly of myriad submicroscopic entities in myriad ever-changing quantum states.† This may at first seem a completely hopeless pretension. For how can we possibly hope to give any account of an assembly that, if it contains just one mole of material, contains no less than 6×10^{23} distinct units? Even a three-body problem defies solution in a completely analytical form; yet we face a 6×10^{23}-body problem. Actually, just *because* of the enormous numbers involved, this problem proves unexpectedly tractable when we give it a *statistical* formulation. From a consideration of assemblies of quantized

---

† Though it enormously simplifies the subsequent analysis, this assumption of quantization is not absolutely essential—and is altogether bypassed in a more laborious but purely classical development.

units, in the next section, we develop three propositions that will prove useful in our statistical analysis. Observe that our concern here is *purely mathematical*, and that we could instead obtain the desired propositions by considering, say, in how many different ways a number of counters can be distributed over the squares of a gameboard.

MICROSTATES AND CONFIGURATIONS

For simplicity, let us consider first an assembly of identical units, localized in space, with permissible quantum states that are associated with an evenly spaced set of energies. An assembly meeting these specifications might be an array of identical one-dimensional harmonic oscillators occupying various fixed positions in a schematic crystal lattice. We stipulate *localization* of the oscillators so that, their identity notwithstanding, each will be rendered distinguishable in principle by its unique geometric placement. We stipulate *identity* of the oscillators so that, in the energy-quantization law $\epsilon_v = (v + \frac{1}{2})hv$, the characteristic frequency v will be the same for any among all the oscillators concerned. The quantum states of any such oscillator can then be depicted as shown in Fig. 1. Since all that concerns us is the *spacing* of these levels, for convenience we have chosen to make our reference zero of energy coincident with the energy of the lowest possible quantum state. That is, for this so-called "ground" state with $v = 0$, we now write $\epsilon_0 \equiv 0$. The energy quantum hv represents the *constant* margin by which each of the higher ("excited") states surpasses in energy the state immediately below it. To bring any oscillator from its ground state to an excited state characterized by some integral value of v, we need only add v quanta with energy hv.

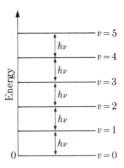

Figure 1

Let us begin with a very simple assembly of three localized oscillators which share three quanta of energy. In how many ways can these three identical quanta be distributed among the three distinguishable oscillators? The ten possible distributions are indicated in Fig. 2—in which the dots are so placed that the letter markings along the abscissa indicate the particular

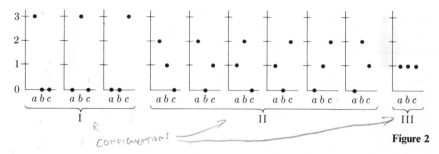

Figure 2

CONFIGURATIONS

oscillator concerned, and the number of energy quanta assigned to it can be read from the ordinate. Each of the ten detailed distributions we call a *microstate,* and it is easy to see that the ten microstates fall in the three groups indicated by Roman numbers. That is, all ten are simply variants of the three basic *configurations* shown in Fig. 3. In configuration I all three energy quanta are assigned to one oscillator, no quanta to the remaining two oscillators, and three microstates develop from this configuration according to whether the three-quantum packet is assigned to oscillator a or to b or to c. In configuration II two quanta are assigned to some one oscillator, one quantum to a second oscillator, no quanta to the third oscillator; and, as indicated in Fig. 2, there are six distinguishable ways in which such assignments can be made. In configuration III one quantum is assigned to each of the three oscillators, and it is evident that there can be but one microstate associated with this configuration.

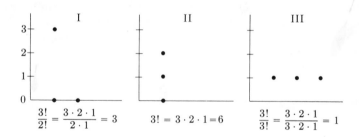

Figure 3

How shall we obtain a systematic count of all the microstates associated with any given configuration? To arrive at the requisite formula, return again to configuration II. Observe that we can assign the first (two-quantum) parcel of energy to any one of three oscillators; having done so, we can assign the second (one-quantum) parcel to either of the two remaining oscillators; there then remains but one oscillator to which we assign the third (nil) parcel. The total number of ways in which the assignments can be made is thus $3 \cdot 2 \cdot 1 = 3!$ (i.e., "three factorial")—which, indeed, duly represents the 6 microstates associated with configuration II. Turning next to configuration I, we have again three choices in assigning the first (three-quantum) parcel, two choices when we assign the second (nil) parcel, and one choice when we assign the third (nil) parcel. But observe that, the last two parcels being the *same*, the final distribution is independent of the *order* in which we assign them. Whether, say, we assign the second parcel to oscillator *b* and the third parcel to oscillator *c*, or *vice versa*, the two verbally distinguishable orders result in precisely the *same* final microstate. That is, $2 \cdot 1 = 2!$ verbally distinguishable assignments collapse into 1 microstate because the two oscillators wind up in the *same* ($v = 0$) quantum level. Hence the total number of microstates associated with configuration I is not 3! but rather $3!/2! = 3$. The same kind of shrinkage of possibilities is seen in even more extreme form in configuration III. Here there is triple occupancy of the same ($v = 1$) quantum level, and the 3! verbally distinguishable assignments collapse into one and the same final microstate. The number of microstates associated with configuration III is then simply $3!/3! = 1$.

By extending this style of analysis, we can easily extract a general formula abundantly useful in more difficult cases. Consider an assembly of some substantial number (N) of localized harmonic oscillators. In how many different ways can we distribute among these oscillators the particular set of energy parcels (including nil parcels) characteristic of the configuration in question? We have N choices of the oscillator to which we assign the first parcel, ($N - 1$) choices in assigning the second, and so on—representing a total of $(N)(N - 1)(N - 2) \cdots (1) = N!$ distinguishable possibilities if no two of the energy parcels are the same. If, on the other hand, some number (η_a) of the parcels are the same, we can obtain only $N!/\eta_a!$ distinct microstates; if η_a of the parcels are of one kind and η_b of some other one kind, we can obtain only $N!/(\eta_a!)(\eta_b!)$ microstates, and so on. The general conclusion is now quite clear. Symbolizing by W the total number of microstates associated with any configuration involving N distinguishable units, we can write:

$$W = \frac{N!}{(\eta_a!)(\eta_b!) \cdots},$$

where η_a represents the number of units assigned the same number of energy quanta (and, hence, occupying the same quantum level), η_b represents the number of units occupying some other one quantum level, ...†

---

† We obtain an identical equation by considering in how many distinct ways a set of counters (e.g., chessmen) can be distributed over an equal number (N) of distinguishable squares. If the counters are all recognizably different from one another, there are $N!$ distinct distributions; but if η_a of the counters are of one kind, and η_b of another, the number of distributions will be reduced to $N!/(\eta_a!)(\eta_b!)$.

The last equation can be represented more compactly as

$$W = \frac{N!}{\prod \eta_n!},$$ (1)

where the symbol $\prod$ instructs us to make a continuing product (even as the symbol $\sum$ instructs us to make a continuing sum) extended over all terms of the form following the symbol, and each of the η_n terms represents the number of units resident in each of the populated quantum levels. Observe that, though we arrived at equation (1) by considering assemblies of harmonic oscillators, with uniform energy spacing between their quantum levels, the actual argument is wholly independent of the supposition of uniformity. *Equation (1) is a general relation*, equally applicable to *any* species of distinguishable unit with *any* energy spacing between its quantum levels. As indicated below, straightforward multiplication of the expanded factorials suffices to establish the number of microstates associated with any configuration for which N is small (<10). For medium-sized values of N (10 to 1000), one can use tabulated values of $N!$ in evaluating W. For very large values of N, we can follow neither of these courses. But, precisely in the limit of large N, an excellent value for $N!$—or, rather, the natural logarithm of $N!$ which we symbolize as $\ln N!$—is supplied by the simplest form of Stirling's approximation,†

$$\ln N! = N \ln N - N.$$ (2)

With equation (1) in hand we can make short work of two additional simple examples. Consider that 5 energy quanta are shared among 5 oscillators. The possible configurations, and the number of microstates associated with each of them, are shown in Fig. 4. Note that even a slight increase in the number of units (and quanta) has produced a sharp increase in the total number of microstates $= \sum W_i = 126$.

As a last example, consider an assembly in which the number of energy quanta is *not* equal to the number of units present: suppose that 5 energy quanta are shared among 10 oscillators. The possible configurations of

---

† Foregoing the elementary application of the calculus that yields a derivation of equation (2), we may just note a crude algebraic argument that offers some rationalization of this form of Stirling's approximation. Consider that the function $N!$ symbolizes the continuing product $(N)(N-1)(N-2)\cdots(2)(1)$. This means a product of N terms with values ranging from a maximum of N to a minimum of 1, and thus averaging about $N/2$. Consequently

$$N! \simeq (N/2)^N$$
$$\ln N! \simeq N \ln (N/2) = N \ln N - N \ln 2 = N \ln N - 0.7 \times N.$$

This proves to be a slight overestimate, which can be much improved merely by replacing the 0.7 by 1.0—and that brings us to equation (2) above.

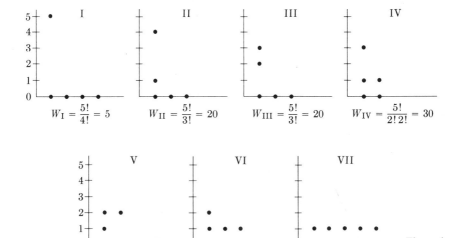

Figure 4

this assembly are easily obtained by adding 5 units to the ground level in each of the configurations shown in Fig. 4—with the results shown in Fig. 5. The calculation of the number of microstates associated with each configuration is given *in extenso*, to call attention to a simple method we will use repeatedly in handling factorial ratios:

$$W_I = \frac{10!}{9!} = \frac{10 \cdot (9!)}{9!} = 10,$$

$$W_{II} = \frac{10!}{8!} = \frac{10 \cdot 9 \cdot (8!)}{8!} = 10 \cdot 9 = 90,$$

$$W_{III} = \frac{10!}{8!} = 90,$$

$$W_{IV} = \frac{10!}{2!7!} = \frac{10 \cdot 9 \cdot 8 \cdot (7!)}{2 \cdot 1 \cdot (7!)} = 10 \cdot 9 \cdot 4 = 360,$$

$$W_V = \frac{10!}{2!7!} = 360,$$

$$W_{VI} = \frac{10!}{3!6!} = \frac{10 \cdot 9 \cdot 8 \cdot 7 \cdot (6!)}{3 \cdot 2 \cdot 1 \cdot (6!)} = 10 \cdot 12 \cdot 7 = 840,$$

$$W_{VII} = \frac{10!}{5!5!} = \frac{10 \cdot 9 \cdot 8 \cdot 7 \cdot 6 \cdot (5!)}{5 \cdot 4 \cdot 3 \cdot 2 \cdot 1 \cdot (5!)} = 6 \cdot 7 \cdot 6 = 252.$$

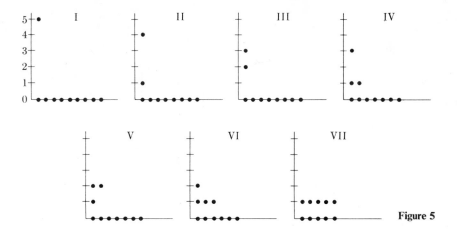

Figure 5

Note that, by doubling the number of units, we have produced close to a ninefold increase in the total number of microstates $= \sum W_i = 2002$.

As the number of units increases further, the total number of microstates skyrockets to unimaginable magnitudes. Thus one can calculate that an assembly of 1000 localized harmonic oscillators sharing 1000 energy quanta possesses more than 10^{600} different microstates. This *is* an unimaginable magnitude: our entire galaxy contains fewer than 10^{70} atoms. Even the estimated total number of atoms in the entire universe is as nothing in comparison with 10^{600}. And though we can offer a compact expression for the total number of microstates that can be assumed by 6×10^{23} oscillators sharing an equal number of energy quanta, that number ($\simeq 10^{10^{23}}$) is essentially meaningless, inconceivably immense.

This explosive expansion of the total number of microstates with increasing N is a direct consequence of the mathematics of permutations, from which arises also a second consequence of no less importance. We can detect the emergence of this further effect in results already obtained. Let us compare our findings for the 5 unit-5 quantum assembly with those for the 10 unit-5 quantum assembly. In Fig. 6 we represent by shaded and open bars respectively the number of microstates associated with each configuration of these two assemblies. If we make the width of each bar equal to one unit of horizontal distance, the numbers of units of area covered by the solid and open bars respectively will indicate the *total* numbers of microstates that can be assumed by the 5-5 and 10-5 assemblies. The ratio of the areas does indeed reflect the approximately $16:1$ value earlier established as the ratio of those numbers. But we have yet to note the most significant feature of the graph: the conspicuous peak representing the number of microstates associated with one configuration (VI) of the 10-5 assembly. Where for the

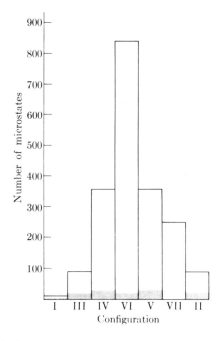

Figure 6

5-5 assembly we find only a modest variation in the number of microstates from one configuration to another, for the 10-5 assembly we observe a clearly predominant configuration with which is associated 42 percent ($= \frac{840}{2002} \times 100$) or close to one-half of the total number of microstates associated with all seven configurations.

Statistical analysis shows that the emergence of a *predominant configuration* is characteristic of any assembly with a large number (N) of units, and that the peaking already observable with a 10-5 assembly becomes ever more pronounced as N increases. We can easily bring out the essence of this analysis by considering the results obtained by repeatedly tossing a well-balanced coin, which can fall in only *two* ways: head (H) or tail (T). If we toss the coin twice, the possible results are then as shown.

$$H \quad H$$
$$H \quad T$$
$$T \quad H$$
$$T \quad T$$

Having stipulated the coin to be well balanced, and assuming that the outcome of the first toss can have no effect on the outcome of the second toss, we may regard all four results as equally probable. Each of these four results we consider a distinct microstate. To them correspond three configurations which reflect the *total numbers* of heads and tails, irrespective of the *order* in which they are obtained. The three configurations are $2H/$, $1H/1T$, and $2T/$. Ordinarily the configurations are what interest us, but it is only the microstates that are equivalent in probability. Thus, of the three configurations, the 1 : 1 possibility is twice as likely to materialize as either of the other configurations—because two microstates are associated with the 1 : 1 configuration, and only one microstate with each of the other two configurations.

Suppose that the coin is flipped a total of 4 times. There are then 16 possible results; i.e., 2^4 microstates, since either of the two possible outcomes of the first toss may combine with either of two possible outcomes of the second toss, and so on. The possible results are as shown.

$$
\begin{array}{llll}
\text{I}\{H\ H\ H\ H &
\begin{array}{l} H\ H\ T\ T \\ H\ T\ H\ T \\ T\ H\ H\ T \\ H\ T\ T\ H \\ T\ H\ T\ H \\ T\ T\ H\ H \end{array}
&
\begin{array}{l} H\ T\ T\ T \\ T\ H\ T\ T \\ T\ T\ H\ T \\ T\ T\ T\ H \end{array}
\end{array}
$$

I{ H H H H

II{ H H H T
H H T H
H T H H
T H H H

III{ H H T T
H T H T
T H H T
H T T H
T H T H
T T H H

IV{ H T T T
T H T T
T T H T
T T T H

V{ T T T T

The 1 : 1 mixture of heads and tails—ultimately the overwhelmingly predominant configuration—has barely begun to stand out. But we can already measure all other configurations relative to this one, by defining a ratio-term (A) as follows:

$$A_X = \frac{\text{Number of microstates associated with configuration } X}{\text{Number of microstates associated with 1:1 mixture}}.$$

The A-ratios in the present case are seen to be as follows.

A_I	A_II	A_III	A_IV	A_V
0.166	0.666	1.000	0.666	0.166

In terms of ratios like these, we can easily follow the progressive development, with increasing number of tosses, of a sharp maximum centered on the predominant configuration. After 6 tosses, there are 7 possible configurations—$6H/$, $5H/1T$, $4H/2T$, $3H/3T$, $2H/4T$, $1H/5T$, $6T/$ —to each of which we assign an index-number defined by the fraction (number of heads/total number of tosses). Again we calculate the A-ratios relative to the predominant 1 : 1 mixture, i.e., the 3 : 3 configuration. Plotting these A-ratios as ordinate *versus* an abscissa representing the corresponding configuration index-number, we obtain the "points" (indicated by tiny triangles) that fall on the $N = 6$ curve in Fig. 7. Actually, only the points themselves are at all meaningful, but a curve has been drawn through them to bring out the

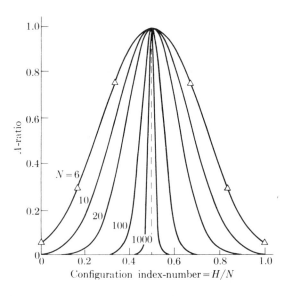

Figure 7

general *shape* of the distribution of microstates among the various possible configurations. On the same set of axes we can plot also the curves that similarly display the situation when the number (N) of tosses is increased to 10, 20, 100, and 1000. The number (W) of microstates associated with each of the ($N + 1$) possible configurations is always calculable. We have only to specify the number of heads (H) and the number of tails (T) that characterize the particular configuration at issue, after which we can easily bring equation (1) to bear, by writing:

$$W = \frac{N!}{H!T!}.$$

With the W's thus calculated, we can go on to calculate A-ratios, relative to the predominant (1 : 1) configuration. Let us then plot the sets of ratios obtained for progressively increasing values of N, as shown in Fig. 7. Observe how an ever-increasing fraction of all possible microstates comes to be associated with one comparatively small group of configurations centered on the predominant configuration. By the time we reach $N = 1000$, an overwhelming proportion of the $2^{1000} \simeq 10^{300}$ microstates then possible are associated with the small set of configurations having H/N values falling in the range 0.45 to 0.55. And by still further increasing N, we can so narrow this range that, at last, it becomes overwhelmingly probable that no actual set of N tosses will yield an H/N value appreciably different from 0.500.

On page 22 we demonstrate that large assemblies of atomic and molecular units display the *same* general behavior. As an example, imagine that we were to use equation (1) to calculate the number of microstates associated with each of the billions of configurations accessible when 1000 quanta are present in an assembly of 1000 harmonic oscillators. Imagine that we were to order these configurations systematically along an abscissa: placing the predominant configuration at the center, we would set the configurations most like it immediately on either side of it, while configurations progressively less and less like the predominant configuration would be set at progressively greater and greater distances from the center. Imagine that, instead of calculating and plotting A-ratios as we did in Fig. 7, we were to construct a bar graph (as we did in Fig. 6) displaying the actual number of microstates associated with each of the configurations. The readily calculable smooth curve that represents the envelope enclosing these billions of bars is sketched in Fig. 8. The area under this curve is of the order of 10^{600}, but the important thing is that almost 100 percent of this area falls under a central peak—the sharpness of which becomes ever more extreme with assemblies of larger and larger numbers of units.

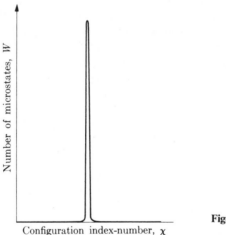

Figure 8

This conclusion sets us looking for some easy way to identify the predominant configuration of a given assembly. We might think to use equation (1) to calculate the number of microstates associated with each possible configuration of the assembly and, from the results so obtained, pick out the predominant configuration. Though always *imaginable* as feasible in principle, this essentially trial-and-error procedure becomes wholly impossible in practice when, even for moderately small systems, the number of configurations to be examined runs into and beyond the billions. Fortunately, a far more elegant and economical approach to the problem becomes apparent as soon as a curve like that shown in Fig. 8 is viewed from the perspective of the differential calculus. We seek a criterion that will identify the predominant configuration for which the number of associated microstates reaches a maximum value. The predominant configuration corresponds to the very peak of the curve, where a tangent line must lie horizontally. Hence the criterion for the predominant configuration is simply that $dW/d\chi = 0$, where $d\chi$ denotes a change from the predominant configuration to another configuration only "infinitesimally" different from it.

At first sight, one may well question the propriety of this style of operation. In passing from one configuration to another, after all, we must always shift irreducibly *integral* quanta of energy; and the consequent alteration in W is always irreducibly *integral*, and often numerically large. Thus the changes denoted by dW and $d\chi$ certainly cannot be regarded as infinitesimal in any *absolute* sense. However, successful application of the differential calculus demands only that changes be infinitesimal in a *relative* sense. This condition can easily be met by stipulating a sufficiently large assembly containing, if you like, 6×10^{23} each of units and quanta—though even 1000 of each would be ample. A shift from one to another of the trillions of trillions of possible configurations requires, we will find, no more than a reassignment of two quanta to two units. But surely the reassignment of 2 (or even 2,000,000) out of 6×10^{23} quanta, to 2 (or even 2,000,000) out of 6×10^{23} units, fully qualifies as infinitesimal in a relative sense. And, though irreducibly integral, surely the consequent change in W—even if it were as large as 10^{10}—fully qualifies as infinitesimal relative to a total value of W that runs to the order of $10^{10^{23}}$. For any sufficiently large assembly, we may then properly regard W as an effectively *continuous* function of the configuration index χ. Hence we need not hesitate to use the criterion $dW/d\chi = 0$ to identify the predominant configuration. In the next section we examine how the use of this criterion leads to a remarkably simple formula that fully describes the predominant configuration.

THE BOLTZMANN DISTRIBUTION LAW

Consider an *isolated macroscopic* assembly of N harmonic oscillators, identical but rendered distinguishable by spatial localization, which share a large number (Q) of those energy quanta each of which suffices to promote an oscillator from one quantum level to the next level above it. By stipulating that the assembly be isolated, we ensure the constancy of both N and Q. We ask now: For which of the enormous number of configurations that can be assumed by this assembly will the number of associated microstates (W) realize its maximum value?

Consider any three *successive* quantum levels, l, m, and n—with progressively increasing energies ϵ_l, ϵ_m, and ϵ_n respectively—populated by numbers of oscillators η_l, η_m, and η_n respectively. Our projected analysis requires that the population numbers be *large*, but we have already ensured this condition by stipulating a macroscopic assembly. Now, for any given configuration, the populations η_l, η_m, and η_n can be represented by the lengths of horizontal lines like those in Fig. 9. For the number of microstates associated with this configuration we can write:

$$W = \frac{N!}{\eta_a! \cdots \eta_l! \eta_m! \eta_n! \cdots \eta_z!},$$

where, as indicated, we must insert in the denominator the appropriate population numbers of *all* the quantum levels that are tenanted in the configuration at issue.

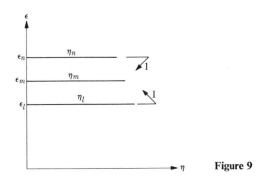

Figure 9

Let us make a minimum change in the initial configuration. Since the assembly has been stipulated to be an isolated one, whatever change we make must hold invariant the value of N and the value of Q. A minimum change consonant with these requirements is the transfer indicated in the figure: into level m we shift one oscillator from each of the levels l and n. Observe that the extra energy quantum required to promote one oscillator from level l to level m is supplied by the energy quantum released when one oscillator drops from level n to level m. Thus, while maintaining strict constancy of both N and Q, we have created a new configuration (distinguished by a prime) that differs from the first only in that:

$$\eta_l' = \eta_l - 1, \qquad \eta_m' = \eta_m + 2, \qquad \eta_n' = \eta_n - 1.$$

For this new configuration we can write:

$$W' = \frac{N!}{\eta_a! \cdots (\eta_l - 1)!(\eta_m + 2)!(\eta_n - 1)! \cdots \eta_z!}.$$

Suppose now that the original configuration were in fact the predominant configuration for which the number of associated microstates reaches its maximum value. In that case $dW/d\chi = 0$; i.e., the relatively infinitesimal change $(d\chi)$ in configuration produces essentially *zero* change in W. To all intents and purposes, therefore, $W = W'$—which means:

$$\frac{N!}{\eta_a! \cdots \eta_l! \eta_m! \eta_n! \cdots \eta_z!} = \frac{N!}{\eta_a! \cdots (\eta_l - 1)!(\eta_m + 2)!(\eta_n - 1)! \cdots \eta_z!}.$$

Canceling the terms common to both fractions, we invert both sides of the equation and obtain

$$\eta_l! \eta_m! \eta_n! = (\eta_l - 1)!(\eta_m + 2)!(\eta_n - 1)!.$$

Minor rearrangement and a slight expansion of some of the factorial terms then yields

$$\frac{\eta_l!}{(\eta_l - 1)!} \cdot \frac{\eta_n!}{(\eta_n - 1)!} = \frac{(\eta_m + 2)!}{\eta_m!},$$

$$\frac{\eta_l(\eta_l - 1)!}{(\eta_l - 1)!} \cdot \frac{\eta_n(\eta_n - 1)!}{(\eta_n - 1)!} = \frac{(\eta_m + 2)(\eta_m + 1)\eta_m!}{\eta_m!}.$$

Cancellation of the factorials now leaves nothing but

$$(\eta_l)(\eta_n) = (\eta_m + 2)(\eta_m + 1).$$

Such numbers as 1 and 2 are clearly negligible in comparison to population numbers all of which were at the outset stipulated to be very large. Hence:

$$(\eta_l)(\eta_n) = (\eta_m)^2 \quad \text{or} \quad \frac{\eta_l}{\eta_m} = \frac{\eta_m}{\eta_n}. \tag{a}$$

Levels l, m, and n are simply any three adjacent levels drawn at random from a great number of possibilities. Had we chosen instead levels k, l, and m, we would have found:

$$\frac{\eta_k}{\eta_l} = \frac{\eta_l}{\eta_m}.$$

Combination of the last two equations then yields what is obviously just one segment of a continuing geometric series:

$$\cdots = \frac{\eta_k}{\eta_l} = \frac{\eta_l}{\eta_m} = \frac{\eta_m}{\eta_n} = \cdots$$

For an isolated macroscopic assembly of harmonic oscillators, with *uniform* energy spacing between their quantum states, this geometric series offers us a remarkably simple description of the predominant configuration. But, turning now from harmonic oscillators, we must still learn to cope with assemblies of other species of units, for which the energy spacing of successive quantum states is *not uniform*. What relation will then describe the sequence of population numbers that makes up the predominant configuration? To answer this question we need only carry through, for the *general* case, the same kind of analysis just made for the special case of harmonic oscillators.

Consider an isolated macroscopic assembly of *any* species of identical but distinguishable units, with *any* kind of energy spacing between their successive quantum states. We focus now on *any* three levels (l, m, and n) with energies $\epsilon_l < \epsilon_m < \epsilon_n$ respectively—as shown in Fig. 10. Let us suppose that between these levels, which need *not* be adjacent, the energy spacings can be adequately approximated by the equation:

$$\frac{\epsilon_n - \epsilon_m}{\epsilon_m - \epsilon_l} = \frac{p}{q}, \tag{b}$$

where p and q are reasonably small† positive integers. Let the original configuration now undergo some slight change which (in conformity with the stipulation of an isolated assembly) holds invariant the total number of units and the total amount of energy. Such a change is the withdrawal of $(p + q)$ units from level m, with transfer of q of them to level n and p of them to level l. That this change maintains constancy of the total energy is easily seen by rewriting the last equation in the form:

$$q(\epsilon_n - \epsilon_m) + p(\epsilon_l - \epsilon_m) = 0.$$

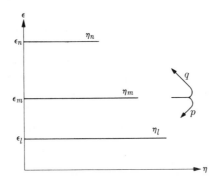

Figure 10

---

† Reasonably small means only small relative to the numbers of units in the levels l, m, and n. If sizable values of p and q are required adequately to express the ratio of the energy spacings, one need only increase the size of the system used—until η_l, η_m, and η_n vastly exceed p and q.

Let us suppose that the original configuration was the predominant configuration for which W takes on its maximum value. In that case $dW/d\chi = 0$, and the indicated rearrangement should leave the value of W substantially unaltered. Proceeding as before, we then write:

$$\frac{N!}{\eta_a!\cdots\eta_l!\eta_m!\eta_n!\cdots\eta_z!} = \frac{N!}{\eta_a!\cdots(\eta_l + p)!(\eta_m - p - q)!(\eta_n + q)!\cdots\eta_z!},$$

$$\frac{(\eta_l + p)!}{\eta_l!}\cdot\frac{(\eta_n + q)!}{\eta_n!} = \frac{\eta_m!}{(\eta_m - p - q)!},$$

$$\frac{(\eta_l + p)\cdots(\eta_l + 1)\eta_l!}{\eta_l!}\cdot\frac{(\eta_n + q)\cdots(\eta_n + 1)\eta_n!}{\eta_n!}$$

$$= \frac{\eta_m(\eta_m - 1)\cdots(\eta_m - p - q + 1)(\eta_m - p - q)!}{(\eta_m - p - q)!}$$

$$(\eta_l + p)\cdots(\eta_l + 1) \times (\eta_n + q)\cdots(\eta_n + 1) = (\eta_m)(\eta_m - 1)\cdots(\eta_m - p - q + 1).$$

In view of the initial stipulation that $\eta_l, \eta_m, \eta_n \gg p, q$, we can properly reduce the last expression to

$$(\eta_l)^p \times (\eta_n)^q = (\eta_m)^{p+q} \qquad \text{or} \qquad \left[\frac{\eta_l}{\eta_m}\right]^p = \left[\frac{\eta_m}{\eta_n}\right]^q. \qquad (c)$$

Observe that in the special case that $p = q$ (i.e., in the event that the energy spacing is uniform, as it is for harmonic oscillators) the general equation just obtained reduces to equation (a), previously derived for an assembly of harmonic oscillators. This concordance of equations (a) and (c) further testifies that these relations are duly independent of the particular small change we subject to analysis. For observe that in obtaining equation (a) we shifted units *into* the middle level, while in obtaining equation (c) we shifted units *out* of the middle level.

Whatever may be the energy spacing between the quantum levels concerned, equation (c) is applicable to any isolated assembly of identical but distinguishable units. However, this equation is not as useful as another equally general relation we now derive by rewriting equation (c) as

$$p \ln\frac{\eta_l}{\eta_m} = q \ln\frac{\eta_m}{\eta_n} \qquad \text{or} \qquad \frac{p}{q}\ln\frac{\eta_l}{\eta_m} = \ln\frac{\eta_m}{\eta_n}.$$

Equation (b) permits us to substitute for p/q, after which minor rearrangement yields

$$\frac{1}{\epsilon_m - \epsilon_l}\ln\frac{\eta_l}{\eta_m} = \frac{1}{\epsilon_n - \epsilon_m}\ln\frac{\eta_m}{\eta_n}.$$

Levels l, m, and n are simply *any* three quantum levels. Had we chosen instead levels k, l, and m we would have found

$$\frac{1}{\epsilon_l - \epsilon_k} \ln \frac{\eta_k}{\eta_l} = \frac{1}{\epsilon_m - \epsilon_l} \ln \frac{\eta_l}{\eta_m}.$$

Hence, *whatever* the two quantum states on which we focus, the indicated function must retain exactly the same value. That is, throughout the set of quantum levels distinctive of the species of units composing the given isolated assembly, this function must be a *constant* we hereafter symbolize as β. For any two quantum states, say i and n, we may then write

$$\frac{1}{\epsilon_i - \epsilon_n} \ln \frac{\eta_n}{\eta_i} = \beta \qquad \text{or} \qquad \ln \frac{\eta_n}{\eta_i} = \beta(\epsilon_i - \epsilon_n).$$

If we take i to symbolize the ground state, with population η_0 and energy $\epsilon_0 \equiv 0$, the last equation reduces to

$$\ln (\eta_n/\eta_0) = -\beta\epsilon_n \tag{3}$$

or

$$\eta_n/\eta_0 = e^{-\beta\epsilon_n}.$$

Here, in the celebrated Boltzmann distribution law, we have at last arrived at the characterization we sought. For an isolated macroscopic assembly of *any* species of identical but distinguishable units, with *any* kind of energy spacing between their quantum states, the predominant configuration has been fully defined by an explicit functional relation between the energy and the population of each quantum level.

Identity of the Significant Configurations. With the aid of the Boltzmann law, we can now produce some solid quantitative evidence for a qualitative proposition merely *asserted* on page 13. Consider an isolated macroscopic assembly of any one species of distinguishable units. Let us discuss just two configurations of this assembly: the predominant configuration in which the number of associated microstates assumes its maximum value W_{max}, and some other very slightly shifted configuration for which the number of microstates is W. Let α_n symbolize the corresponding *fractional* alteration in the number of units present in the nth quantum level. That is, marking with a prime the population numbers in the shifted configuration, we write

$$\alpha_n \equiv \frac{\eta_n' - \eta_n}{\eta_n}.$$

For any level in which the population increases, α will be positive; for any level in which the population decreases, α will be negative. And, since the shifted configuration has been supposed only minutely different from the predominant configuration, in all cases $|\alpha| \ll 1$.

By rearrangement of the definition of α_n, we can easily express the actual change in the population of the nth quantum level as

$$\eta_n' - \eta_n = \alpha_n \eta_n.$$

In the isolated system both the total number of units (N) and the total energy (E) must remain strictly constant. All the changes in population numbers must then sum to a net value (ΔN) that is zero, so that

$$\Delta N = (\eta_0' - \eta_0) + (\eta_1' - \eta_1) + \cdots = \sum \alpha_n \eta_n = 0. \tag{d}$$

Moreover, since all the changes in energy consequent to the shifts in population numbers must also sum to a net value (ΔE) that is zero,

$$\Delta E = \epsilon_0(\eta_0' - \eta_0) + \epsilon_1(\eta_1' - \eta_1) + \cdots = \sum \epsilon_n \alpha_n \eta_n = 0. \tag{e}$$

These two restrictive conditions will play an important role later in the analysis, which now becomes somewhat tedious. Readers who wish to avoid the tedium may skip directly to the passage indicated by the marginal arrow on the facing page.

In terms of α_n, the population of the nth quantum level in the shifted configuration can be expressed as

$$\eta_n' = \eta_n + \alpha_n \eta_n.$$

As the ratio of W_{max} to W in the shifted configuration, equation (1) at once supplies:

$$\frac{W_{max}}{W} = \frac{N\,!/\prod \eta_n!}{N\,!/\prod(\eta_n + \alpha_n \eta_n)!} = \frac{\prod(\eta_n + \alpha_n \eta_n)!}{\prod \eta_n!}.$$

Since we can express the logarithm of the continuing products as continuing sums of logarithmic terms,

$$\ln (W_{max}/W) = \sum \ln (\eta_n + \alpha_n \eta_n)! - \sum \ln \eta_n!.$$

In this macroscopic assembly all the η_n-terms will be large enough to justify application of Stirling's approximation, which yields:

$$\ln (W_{max}/W) = \sum (\eta_n + \alpha_n \eta_n) \ln (\eta_n + \alpha_n \eta_n) - \sum (\eta_n + \alpha_n \eta_n)$$
$$- \sum \eta_n \ln \eta_n + \sum \eta_n.$$

Canceling $\sum \eta_n$ between the second and the fourth terms on the right, we progressively separate the first term on the right

$$\ln (W_{max}/W) = \sum \eta_n \ln \eta_n(1 + \alpha_n) + \sum \alpha_n\eta_n \ln \eta_n(1 + \alpha_n)$$
$$- \sum \alpha_n\eta_n - \sum \eta_n \ln \eta_n$$
$$= \sum \eta_n \ln \eta_n + \sum \eta_n \ln (1 + \alpha_n) + \sum \alpha_n\eta_n \ln \eta_n$$
$$+ \sum \alpha_n\eta_n \ln (1 + \alpha_n) - \sum \alpha_n\eta_n - \sum \eta_n \ln \eta_n.$$

The first and last terms on the right now cancel. And with $|\alpha| \ll 1$, to an excellent degree of approximation $\ln (1 + \alpha) \simeq \alpha$, for both positive and negative values of α. Therefore

$$\ln (W_{max}/W) = \sum \alpha_n\eta_n + \sum \alpha_n\eta_n \ln \eta_n + \sum \eta_n\alpha_n^2 - \sum \alpha_n\eta_n$$
$$= \sum \alpha_n\eta_n \ln \eta_n + \sum \eta_n\alpha_n^2 .$$

So far we have come on the strength of nothing but equation (1) and a reliable pair of purely mathematical approximations dependent only on the largeness of the η's and the smallness of the α's. Calling now on the Boltzmann distribution law, we rewrite equation (3) in the form

$$\ln \eta_n = \ln \eta_0 - \beta \epsilon_n.$$

Substitution in the last equation for $\ln (W_{max}/W)$ yields

$$\ln (W_{max}/W) = \sum \alpha_n\eta_n \ln \eta_0 - \sum \beta\epsilon_n\alpha_n\eta_n + \sum \eta_n\alpha_n^2$$
$$= (\ln \eta_0) \sum \alpha_n\eta_n - \beta \sum \epsilon_n\alpha_n\eta_n + \sum \eta_n\alpha_n^2$$

since, as regards these summations, both $\ln \eta_0$ and β are *constants* characteristic of the original configuration. Now equations (d) and (e) indicate that, for this *isolated* system, the requirements $\Delta N = 0$ and $\Delta E = 0$ ensure that both of the first two summations vanish identically. All that then remains is

$$\ln (W_{max}/W) = \sum \eta_n\alpha_n^2 .$$

What one average value will best represent the extent to which the shifted configuration differs from the predominant configuration? The simple weighted average $\sum (\eta_n\alpha_n)/N$ is wholly insufficient: as just noted, the requirement $\Delta N = 0$ will in every case reduce this summation to zero. However, because all terms are then squared, both positive and negative values of α *will* be duly reflected in the root-mean-square average value defined by writing

$$\bar{\alpha} \equiv \left[\frac{\sum \eta_n\alpha_n^2}{N} \right]^{1/2} .$$

This formulation demands that the value of α^2 for each quantum level be multiplied by a weighting factor which, appropriately enough, is simply the number of units originally resident in that level. Division by N then yields a weighted-average value of α^2, and $\bar{\alpha}$ is then nothing but the square root of this average value of α^2. Now on squaring and rearranging the equation that defines $\bar{\alpha}$, we find

$$N\bar{\alpha}^2 = \sum \eta_n \alpha_n^2 .$$

Substitution in the last equation of the preceding paragraph then yields

$$\ln (W_{max}/W) = N\bar{\alpha}^2$$

$$W/W_{max} = e^{-N\bar{\alpha}^2}. \tag{4}$$

We now have in $\bar{\alpha}$, as we never had in χ, a fully quantitative measure of configuration change. To discover how W varies with $\bar{\alpha}$, we consider an isolated macroscopic assembly that contains just one-sixth of a mole of units, so that $N = 10^{23}$. Let this assembly be shifted from its predominant configuration to another configuration so nearly identical that the average change in population numbers is just one ten-billionth of their original values. As the change in W consequent to this minute shift, to a configuration with $\bar{\alpha} = 10^{-10}$, equation (4) at once yields

$$\frac{W}{W_{max}} = e^{-N\bar{\alpha}^2} = e^{-10^{23} \cdot 10^{-20}} = e^{-1000} = 10^{-434} .$$

Thus the shift, from the predominant configuration to a new configuration apparently only trivially different from it, has produced an almost incredible collapse of the number of associated microstates, from W_{max} to $W_{max} \times 10^{-434}$. For something of the order of a mole of units, the peak on a plot of W *versus* χ is thus *extremely* narrow—far narrower than the peak sketched for a much smaller assembly in Fig. 8. Even though a shift having the value $\bar{\alpha} = 10^{-10}$ is far too minute to be reflected in any perceptible change in the macroscopic properties of this assembly, such a shift amply suffices to produce a new configuration that effectively contributes *nothing* to the total number of microstates accessible to the assembly. And so we come to this conclusion:

Of the immense total number of microstates that can be assumed by a *large* assembly, an overwhelming proportion arises from one comparatively small set of configurations centered on, and only minutely different from, the predominant configuration—with which they share an empirically identical set of macroscopic properties.

We shall refer to this notable set of configurations as PG (i.e., predominant-group) configurations.

PHYSICAL MEANING OF THE DISTRIBUTION LAW

From the almost purely mathematical inquiry conducted in the preceding sections, we have extracted a general distribution law. What is the physical significance of the predominant configuration described by that law? To answer this question, we now introduce the basic postulate of statistical mechanics:

Over long periods of time, every possible microstate of an isolated assembly occurs with equal probability.

The concept of probability is a slippery one, but the meaning of this postulate can be made quite clear by relating it to an imaginable experiment. Suppose we were able to observe the microstates successively prevailing in an isolated assembly assumed to be unperturbed by our observations. As the number of those observations is increased without limit, the postulate asserts, we will find that the *ratio* of the numbers of observed occurrences of any two microstates will approach *unity*. This statement is intended only to clarify the meaning of the postulate, *not* to describe a realizable experiment. The experiment is unrealizable because, in the first place, the microstates that are so distinct conceptually are totally indeterminate empirically. To determine the microstate of an assembly containing one mole of units, we would have to measure the precise quantum state of each of 6×10^{23} units. All these measurements would have to be made essentially instantaneously because, of course, the quantum states of the various units change incessantly *and rapidly*. Even supposing that we were able to observe complete microstates, and could do so at the rate of 10^6 microstates per second, we would still face a second difficulty: the entire estimated lifetime of the universe as we know it would be insufficient to permit observation of as much as one occurrence of each of the 10^{600} different microstates possible in a tiny assembly of 1000 harmonic oscillators sharing 1000 energy quanta.

We cannot then defend our basic postulate as an empirically verifiable generalization or law. And, though one can cite many *a priori* arguments in support of that postulate, such arguments can never induce complete conviction. But neither the empirical demonstration nor the arguments are needed if we are content to treat the basic postulate *as* a postulate. For consider, as another example, why we accredit Newton's first postulate, the principle of inertia. We do so *not* because it is empirically verifiable, *not* because *a priori* arguments can be cited in its favor. We accredit the Newtonian postulate simply because we can deduce from it (in association with the other Newtonian postulates) an enormous number of consequences that are empirically verifiable, and exhaustively verified. In exactly the same way, the basic postulate of statistical mechanics—though not itself experimentally verifiable—is accredited by the successful verification of an enormous number of empirical predictions obtained by deduction from that postulate.

We accept the basic postulate of statistical mechanics (as we accept Newton's first postulate) because statistical mechanics (like Newtonian dynamics) is a science that deals successfully with the physical world of our experience.

To display the general thrust of the apparently innocuous basic postulate, we now consider an assembly that is (1) *isolated*, so that the total energy and the number of units remain constant; and (2) *macroscopic*, so that the number of units is large, the number of possible configurations immense, and the total number of possible microstates astronomical. We stipulate further that the assembly is to be constituted of units that are (3) *identical* but *distinguishable*, as are the units in our familiar set of localized harmonic oscillators. Lastly, we stipulate that the units are to be (4) *weakly coupled*. By invoking "coupling"—the possibility of energy exchanges among the units of the assembly—we ensure that each microstate and configuration of the assembly will be rapidly succeeded by other microstates and configurations. By decreeing the coupling to be "weak," we ensure that each unit preserves the array of quantum states it would have in isolation, substantially unperturbed by the presence of the other units with which it is coupled.

Having devised an assembly conformable with the four stipulations just noted, let us allow it to "settle" for a time sufficient to permit even weak coupling completely to dissipate the original configuration. If we then set ourselves to long-continued close observation of the macroscopic state of this assembly, what will we observe? This question we can answer on the basis of the decisive argument presented in the following numbered steps.

1. Our basic postulate asserts that, whenever the assembly is observed, it is equally likely to be found in any of its myriad possible microstates.

2. In a macroscopic assembly like ours, an overwhelming proportion of all possible microstates arises from one relatively small set of PG configurations.

3. Points 1 and 2 together entail the following conclusion: with overwhelming probability, on observation the assembly will be found in some PG configuration.

4. Too nearly identical to be differentiated by any detectable difference in the macroscopic properties of the assembly, all the PG configurations must appear to actual observation as one and the same macroscopic state of the assembly.

5. Points 3 and 4 together entail the following conclusion: with overwhelming probability, on observation the assembly will be found always in the *one* macroscopic state that corresponds to all the PG configurations typified by the predominant configuration.

The question with which we began the paragraph is thus answered, and let us pause a moment to reflect on this interim conclusion. Having stipulated weak coupling of the units, we are assured that the assembly is potentially capable of assuming *any* of its configurations—not just PG configurations. Many of these other configurations differ sufficiently from any PG configuration to appear as radically different macroscopic states of the assembly. Hence the stipulation of weak coupling ensures that the assembly *could* show a change of macroscopic state. Yet with overwhelming probability we find it always in the *same* macroscopic state. What do we say when, over a long period of time, we find *unchanging* the macroscopic state of an isolated system we know to be *capable of change*? We say:

6. The unchanging state of an isolated macroscopic system is to be called its *equilibrium state*.

7. Points 5 and 6 together entail the following conclusion: The equilibrium state of an isolated macroscopic assembly is that corresponding to the PG configurations typified by the predominant configuration. In other words:

   At equilibrium, the configuration of an isolated macroscopic assembly is typically that described by the Boltzmann distribution law.

Our abstract analysis here comes down to earth with an immensely powerful general criterion of equilibrium.

Though every microstate accessible to the given assembly occurs with equal probability, observe that this does *not* mean that all its configurations have an equal probability of occurrence. Quite the contrary! If one configuration (P) is associated with some number of microstates (W_P), and a second configuration (Q) is associated with some vastly larger number of microstates (W_Q), configuration Q is far more likely to turn up. Indeed, when the ratio W_Q/W_P is very large, the balance of probability represents a virtual certainty. Thus if the initial configuration of a weakly coupled assembly is one for which W is small, we may confidently expect a "spontaneous" change to configurations for which W is greater. Moreover, this shift must continue until, at last, the assembly attains the PG configurations that together comprehend an overwhelming proportion of all microstates accessible to any large system. And, having once attained PG configuration, the assembly will thereafter be found, with overwhelming probability, in the empirically unchanging equilibrium state characteristic of all those macroscopically identical configurations typically described by the Boltzmann law.

Readily interpretable as a progression toward configurations of maximum W, the equilibration of an isolated assembly may also be interpreted as a progression toward configurations of maximum "disorder." Actually, this second interpretation follows at once from the first. For consider that those configurations we single out as "ordered" are precisely those which, with relatively small values of W, are *least* likely to arise by random chance. As a mundane illustration, imagine a box in which 1000 distinguishable coins all lie heads up. This perfectly ordered configuration can arise in but one way. Suppose that, after shaking the box slightly, we find its contents in the less orderly $900H/100T$ configuration, which can arise in some 10^{140} different ways. Once the box has been shaken, this configuration is favored by a margin of $10^{140}:1$ over the initial perfectly ordered configuration. Suppose then that, after further shaking, we find the contents of the box in the still less ordered $700H/300T$ configuration, which can arise in some 10^{265} different ways. The appearance of this configuration is favored by a margin of $10^{125}:1$ over retention of the intermediate $900H/100T$ configuration, and by a margin of $10^{265}:1$ over reappearance of the initial completely ordered configuration. Suppose finally that the box is subjected to prolonged and very vigorous shaking. We are then most likely to find its contents in some close approximation to the completely random $500H/500T$ configuration, which can arise in some 10^{300} different ways. And now still further shaking is unlikely to produce any substantial departure from the wholly disordered PG configurations in which W has at last assumed its maximum value. This purely pedestrian illustration thus shows how a progressive increase in "disorder" necessarily accompanies an approach to equilibrium characterized by the assumption of configurations with ever-increasing values of W. And what may at first appear to be a purposeful "drive," toward states of maximal disorder, can now be seen to arise simply from the operation of blind chance in an assembly where all microstates remain equally probable, but where an overwhelming proportion of all microstates is associated with the maximally disordered PG configurations.

Whatever interpretation we may choose to place upon it, our conclusion remains this: in an isolated macroscopic assembly, the equilibrium configuration is typically that described by the distribution law $\eta_n = \eta_0 e^{-\beta \epsilon_n}$. Of course this description remains seriously incomplete for so long as β remains an unidentified parameter. What is the physical significance of β? The next section prepares the way for an answer to that question.

VARIATION OF W WITH E

When the energy content of a given assembly is very small, its constituent units must cluster densely in the lower-energy quantum states. Even for the predominant configuration, the value of W must in these circumstances be comparatively small. (Observe that when *all* units repose in the ground state, $W = N!/\prod\eta_n! = N!/\eta_0! = N!/N! = 1$.) When, on the other hand, the same assembly possesses some higher energy, its constituent units will distribute themselves more sparsely over a greater number of quantum levels. Under these conditions the function $\prod \eta_n!$ becomes smaller, and W will be correspondingly increased. Thus the value of W *should* be related to the internal energy (E) of the given assembly. What, precisely, is the operative relation?

Consider a macroscopic body consisting of any one species of loosely coupled distinguishable units. Let the energy of the body be very slightly increased by the input of a minute amount of "heat"—which we may define for the present purpose simply as energy transferred by thermal conduction or radiation. As a consequence of this infinitesimal energy transfer, the units are shifted from their original predominant configuration to a new equilibrium configuration only infinitesimally different. The initial value of W we can calculate from equation (1), $W = N!/\prod \eta_n!$, where the η's are determined by the Boltzmann distribution law. By differentiation, we may then hope to get at the change of W that accompanies the small shift of configuration.

In logarithmic form, equation (1) becomes:

$$\ln W = \ln N! - \ln \prod \eta_n!$$
$$= \ln N! - \sum \ln \eta_n!,$$

where the logarithm of the continuing product has been replaced by the continuing sum of logarithmic terms. With the number of units (N) a *constant* characteristic of the given body, differentiation will now yield

$$d \ln W = -\sum d \ln \eta_n!.$$

For a macroscopic assembly we can use Stirling's approximation—our equation (2)—to obtain the differential of each $\ln \eta!$ term

$$d \ln \eta! = d(\eta \ln \eta - \eta) = \eta \frac{d\eta}{\eta} + \ln \eta \, d\eta - d\eta = \ln \eta \, d\eta.$$

Substitution in the expression for $d \ln W$ then produces

$$d \ln W = -\sum \ln \eta_n \, d\eta_n.$$

For the predominant configuration of the body in its initial equilibrium state, equation (3) offers

$$\ln \eta_n = \ln \eta_0 - \beta \epsilon_n.$$

Substituting for $\ln \eta_n$ in the last equation of the preceding paragraph, we find

$$d \ln W = -\sum (\ln \eta_0) \, d\eta_n + \sum \beta \epsilon_n \, d\eta_n$$
$$= -(\ln \eta_0) \sum d\eta_n + \beta \sum \epsilon_n \, d\eta_n,$$

since, insofar as these summations are concerned, both $\ln \eta_0$ and β are constants characteristic of the original configuration. Evaluation of the two summations is now very easy: indeed, two closely related summations have already been considered on page 20.

Adding the numbers of units present in all the quantum levels concerned, we must obtain the *constant* total number (N) of units present in the given body. Hence

$$\eta_0 + \eta_1 + \eta_2 + \cdots = \sum \eta_n = N = \text{constant}$$

and differentiation yields

$$\sum d\eta_n = dN = 0.$$

Turning to the second summation, we observe that the total energy (E) of the assembly is simply the sum of the energies possessed by the individual units as they stand in the various quantum levels. Consequently

$$\eta_0 \epsilon_0 + \eta_1 \epsilon_1 + \eta_2 \epsilon_2 + \cdots = \sum \eta_n \epsilon_n = E.$$

In circumstances soon to be noted, a small energy transfer may be accomplished with *no* change in the energies of the various quantum levels. And with all the ϵ's constant, differentiation will yield

$$\sum \epsilon_n \, d\eta_n = dE,$$

which is only to say that, in the indicated circumstances, the energy of the assembly can change solely by a redistribution of its units over quantum states with invariant energies. And having now found equivalents for both the summations appearing in the last equation of the preceding paragraph, by substitution we obtain:

$$d \ln W = \beta \, dE. \tag{5}$$

This is just what we sought. While at the outset we could suggest nothing beyond the qualitative intuition that W must increase with E, the relation here derived offers a quantitative measure of the minute change in $\ln W$ produced by a minute change in E. Observe however that this derivation has taken as premise an equation (1) that is valid only for assemblies of distinguishable units. Hence, though more generally demonstrable in the long term, equation (5) is at this point established only in application to assemblies of *distinguishable* units. And, quite apart from this transient restriction on the *kind of assembly*, three continuing restrictions limit the *kind of change* for which equation (5) will suffice.

The first restriction arises from our assumption that, by a transfer of heat, we can increase the total energy of the system without *any* concomitant shift of the energies associated with the quantum states characteristic of the species of units concerned. With units like harmonic oscillators, for which the associated energies are established by the equation $\epsilon_v = (v + \tfrac{1}{2})h\nu$, this assumption seems quite secure. Having no reason to suppose that the species-specific frequency v will vary significantly with modest variations in the energy content of the body, we have every reason to suppose that (with h a universal constant) the energies of the various quantum states will remain effectively constant. But with other species of units the situation is not so simple. Thus in an assembly of gas molecules in free translational motion, the energies (ϵ_r) associated with the various translational quantum states are defined by an equation of the form $\epsilon_r = f(1/V)^{2/3}$, where V symbolizes the volume of the assembly. In such a case we are obviously well advised to require that equation (5) shall be applied only to heat transfers made under *constant-volume* conditions.†

The second restriction is a clearly essential stipulation that the energy transfer shall induce *no chemical or physical changes* capable of altering either the number or the nature of the units present. We must insist on invariance of their total number because the constancy of N was presumed in our derivation of equation (5). And only by insisting that the nature of the units shall remain unchanged can we guarantee that the characteristic energies (ϵ_n) will retain the strict constancy also presumed by our derivation.

The third restriction arises from the step in which we wrote $\sum \beta \epsilon_n \, d\eta_n = \beta \sum \epsilon_n \, d\eta_n$. This rearrangement will be permissible only if β can be regarded as effectively *constant* throughout all the population changes consequent to the transfer of thermal energy. But, with the progressive shift of population numbers linked by the equation $\eta_n/\eta_0 = e^{-\beta \epsilon_n}$, some change of β will necessarily accompany the transfer of any considerable amount of energy. We must then stipulate that equation (5) shall be applied only when the energy transfer dE is effectively *infinitesimal*.

THE APPROACH TO THERMAL EQUILIBRIUM

A straightforward exploitation of equation (5) will now yield us a highly suggestive relation involving nothing but the still enigmatic parameter β. Consider two macroscopic bodies, X and Y, each consisting of any one species of loosely coupled distinguishable units. Let the two bodies be brought into thermal contact, and so insulated that the united system is completely isolated from the rest of the world. How many microstates will be accessible in a given configuration of the united system? Observe that each microstate accessible to the united system necessarily represents some particular combination of one of the microstates accessible to body X with one of the microstates accessible to body Y. Hence the total number of microstates associated with the given configuration of the united system is

---

† Strictly speaking, not only the volume but every other external parameter of the assembly must also remain fixed—a condition that effectively precludes any transfer of energy *save* in the form of heat.

determined by the number of distinct ways in which one of the microstates accessible in X can be combined with one of the microstates accessible in Y. Since any one of the W_X microstates accessible in X may combine with any one of the W_Y microstates accessible in Y, clearly the *product* $W_X \cdot W_Y$ must represent the total number of microstates associated with the given configuration of the united system. A product $W_X \cdot W_Y$ must indeed represent the number of microstates associated with *any* one configuration of the united system, but of course different values for W_X and W_Y will be appropriate to different configurations. That is, as new configurations of the united system arise from redistribution of energy between X and Y, new values of W_X and W_Y will be required properly to represent the numbers of microstates accessible in each body individually.

Let W_{Xi} and W_{Yi} symbolize the numbers of microstates associated with the predominant configurations of the two *separate* bodies, in their initial states of internal equilibrium. Now consider the situation in the first instant after the two bodies have been joined. For this moment, before thermal contact can yield any redistribution of energy, the initial configuration of the united system will be associated with a total number of microstates given by the product $W_{Xi} \cdot W_{Yi}$. However immense this number may be, the initial configuration will be unstable if—through redistribution of its energy—the united system can attain a configuration associated with some much larger value of the product $W_X \cdot W_Y$. And the redistribution must continue until, at last, the system arrives at that *equilibrium* state in which the product $W_X \cdot W_Y$ attains its maximum value. If in their original (separated) condition the two bodies were distinguished by different values of β, what relation of the final values of β will characterize X and Y when they have come to thermal equilibrium? Under conditions that ensure conformity with the three restrictions noted at the end of the last section, application of equation (5) at once supplies an illuminating answer to this question.

For both the approach to equilibrium and the eventual condition of equilibrium, a very compact analytical expression is

$$d(W_X \cdot W_Y) \geq 0.$$

The inequality applies to the approach to equilibrium, during which any net transfer of energy is accompanied by some net increase of the product $W_X \cdot W_Y$. And the equality applies to the condition at equilibrium when, the product $W_X \cdot W_Y$ having attained its maximum value, an infinitesimal energy transfer will produce *no* further change in that product. Actually performing the indicated differentiation, we find

$$W_Y \, dW_X + W_X \, dW_Y \geq 0.$$

Division by $W_X \cdot W_Y$ now yields

$$\frac{dW_X}{W_X} + \frac{dW_Y}{W_Y} \geq 0.$$

Since, for any function f, $d \ln f = df/f$, we may also write

$$d \ln W_X + d \ln W_Y \geq 0.$$

This result we might have obtained even more easily—by reflecting that whenever $W_X \cdot W_Y$ increases, so also must $\ln W_X \cdot W_Y$ increase; and whenever $W_X \cdot W_Y$ reaches its maximum, so also must $\ln W_X \cdot W_Y$ reach *its* maximum. Thus, instead of the first equation in the preceding paragraph, we might as well have written $d \ln W_X \cdot W_Y \geq 0$, which then at once yields the last equation of the preceding paragraph. However derived, in this equation we now substitute on the strength of equation (5), which gives

$$\beta_X \, dE_X + \beta_Y \, dE_Y \geq 0.$$

Having specified that X and Y shall come to equilibrium with each other in isolation from the rest of the world, we know that their total energy must be a *constant*. Hence

$$E_X + E_Y = \text{constant}$$

so that

$$dE_X + dE_Y = 0 \qquad \text{or} \qquad dE_Y = -dE_X.$$

That is, the energy content of Y can increase only at the expense of an equivalent decrease in the energy content of X. Substituting then in the last equation of the preceding paragraph, we find

$$\beta_X \, dE_X - \beta_Y \, dE_X \geq 0.$$

Let us suppose that the approach to equilibrium is characterized by a passage of heat from body Y to body X. In that event dE_X will be a *positive* quantity, cancellation of which reduces the last relation to

$$\beta_X \geq \beta_Y.$$

What is the import of this remarkably simple conclusion?

Referring to an approach to equilibrium we supposed to be characterized by a spontaneous flow of heat from Y to X, the inequality $\beta_X > \beta_Y$ expresses the essential condition for that transfer. Moreover, the rationale of this condition is perfectly evident. With dE_X positive, the transfer of this energy parcel decreases $\ln W_Y$ to the extent $-\beta_Y \, dE_X$, even as it increases $\ln W_X$ to the extent $+\beta_X \, dE_X$. As long as $\beta_X > \beta_Y$, the energy transfer proceeds spontaneously simply because it yields a *net* increase in $\ln W_X \cdot W_Y$. That is, as long as $\beta_X > \beta_Y$, energy passes from Y to X because

$$d \ln W_X \cdot W_Y = d \ln W_X + d \ln W_Y = (\beta_X - \beta_Y) \, dE_X \geq 0.$$

Only on the attainment of thermal equilibrium is the progressive increase in $\ln W_X \cdot W_Y$ finally arrested—at the point where any further heat transfer would produce an increase in $\ln W_X$ that is fully compensated by the accompanying decrease in $\ln W_Y$. And the last relation demonstrates that this condition of equilibrium, to which the equality refers, is understandably attained precisely when $\beta_X = \beta_Y$.

Physical meaning of β. Using any self-consistent scale of temperature (t), we discover empirically that spontaneous transfers of heat from body Y to body X occur only when the temperature of Y exceeds that of X; and, further, that the two bodies stand in thermal equilibrium only when their temperatures have at last become equal. Regardless of the size and composition of the bodies concerned, both the approach to equilibrium and the condition of equilibrium are thus found to be determined *solely* by the following bimodal relation of the temperatures:

$$t_X \leq t_Y.$$

On the other hand, our entirely independent theoretical analysis has established that, regardless of the size and composition of the bodies concerned, both the approach to equilibrium and the condition of equilibrium are determined *solely* by the following bimodal relation of β's:

$$\beta_X \geq \beta_Y.$$

We are thus led to a major conclusion: β must represent some function of temperature and, more specifically, an *inverse* function of temperature. Since *all* bodies in thermal equilibrium possess the same value of β, we may hope to define in terms of β a scale of temperature (θ) that is "absolute" in the sense that it depends on the properties of no one arbitrarily selected thermometric substance. Readings on this scale we denote as °K, in recognition of the work of Lord Kelvin, who was the first fully to grasp the possibility of an absolute temperature scale.

How shall we link β with θ? Considering their inverse relationship, our first thought may be to write $\beta = 1/\theta$, but this formulation is not quite sufficient dimensionally. The Boltzmann distribution law requires that $\eta_n/\eta_0 = e^{-\beta\epsilon_n}$. On the left-hand side appears a dimensionless ratio of population numbers. From this it follows that the right-hand side of the equation—and, more particularly, the exponent—must also be dimensionless. If then the term $\beta\epsilon_n$ is to be dimensionless, we see that β must have the units (energy)$^{-1}$. But, in the relation $\beta = 1/\theta$, β would instead have the units (°K)$^{-1}$. Thus we are led to revise that tentative relation, by writing

$$\beta = 1/k\theta,$$

where k is some constant with units (energy/°K). For bodies standing in thermal equilibrium, the equality of θ's demanded by our concept of temperature will be reconcilable with the demonstrated equality of β's only if we regard k as a "universal" constant—the same for all bodies.

The last equation offers a dimensionally acceptable expression that makes β an inverse function of temperature on the scale defined by writing $\theta \equiv 1/k\beta$. But how can our usual thermometric measurements be expressed as values on this absolute θ-scale? We observe that, though unattainable, $0°$ on the familiar ideal-gas scale (T) is a temperature now known to be indefinitely approachable. Thus $0°T$ appears to represent an *ideal limit* properly qualified to serve as an "absolute zero" of temperature. In this limit a system would be completely drained of all the energy removable from it, and all its component units must then fall together in the ground state. What will be the value of β in these circumstances? If all quantum states above the ground state are to remain untenanted, then of course

$$\eta_n/\eta_0 = e^{-\beta\epsilon_n} \to 0.$$

With $\epsilon_0 \equiv 0$, all other ϵ_n's will be finite positive numbers. Hence the indicated limiting value will be approachable only if $\beta \to \infty$ in the limit of $0°T$. Thus the absolute θ-scale of temperature must coincide with the ideal-gas T-scale in the limit of $0°K$.

We shall defer to p. 67 a demonstration that, by an appropriate choice of the constant k, the θ-scale can be made *everywhere* coincident with ideal-gas T-scale. In anticipation of this demonstration, we shall feel free to write

$$\beta = 1/kT, \tag{6}$$

and T also we shall express in $°K$.

THE CONCEPT OF ENTROPY

Consideration of one last fundamental thermodynamic variable is prompted by a recollection of how the behavior of W yielded us the decisive predictive criteria $dW \geq 0$. That is, for any isolated assembly, we can always predict the direction of spontaneous change as that in which W increases, and the condition of equilibrium as that in which W assumes its maximum attainable value.

No other characteristic of an assembly is as illuminating as W. The more familiar energy concept certainly offers no comparable indications of spontaneity and equilibrium: in an isolated assembly, the total energy must remain *constant* throughout any possible change. For the special case of thermal equilibrium, to be sure, the temperature concept can supply such indications: the spontaneous internal redistribution of energy always proceeds in the direction that yields a reduction of the temperature difference between the two subassemblies, and equilibrium is attained only when both stand at the same temperature. However, though here quite sufficient, the temperature concept is not *generally* competent to supply the criteria of spontaneity and equilibrium we draw from consideration of W. For consider that we aspire to a determination of the direction of spontaneous chemical change, and the condition of chemical equilibrium, in an isolated system throughout which the temperature is no more uniform at the end than at the very beginning. In such a case we surely *cannot* draw from the temperature concept, or from the energy concept, any of the indications again readily drawn from the behavior of W. Saying nothing for the moment about how W is to be evaluated for such systems of changing composition, we confidently expect that the direction of spontaneous chemical change must *still* be that yielding some net increase of W; and the condition of equilibrium *still* that in which W assumes the maximum value accessible to the isolated system as a whole.

In what terms may we best conduct our consideration of W? For any macroscopic assembly, the values of W will be inconveniently immense; and we shall adopt the simple expedient of discussing not W itself but, rather, $\ln W$. We lose nothing by adopting this expedient: when W increases, so also must $\ln W$; when W reaches its maximum value, so also must $\ln W$. And by using $\ln W$ instead of W itself, we easily reduce the relevant numbers to more manageable magnitudes. (Consider that 10^{23} is a huge number, while $\ln 10^{23}$ is just 53.) But a far more fundamental reason also motivates our decision to work with $\ln W$ rather than with W.

If we define a system parameter $\mathscr{A}$ by writing $\mathscr{A} \equiv \ln W$, we will have created in $\mathscr{A}$ a simple "extensive" property of the system concerned. What does that mean? Consider two identical systems (A and B) which we unite to form a single double-size system (AB). An *intensive* property, like temperature, remains the same in the double-size system AB as it is in A and B individually. On the other hand, an *extensive* property, like energy, will be twice as great in the double-size system AB as it is in A and B individually (i.e., $E_{AB} = E_A + E_B$). To see that, like energy, the parameter $\mathscr{A}$ represents an extensive property, we have only to recall our earlier conclusion that in just these circumstances $W_{AB} = W_A \cdot W_B$. Consequently

$$\ln W_{AB} = \ln W_A + \ln W_B.$$

And if by definition $\mathscr{A}_A \equiv \ln W_A$, $\mathscr{A}_B \equiv \ln W_B$, and $\mathscr{A}_{AB} \equiv \ln W_{AB}$, then it follows at once that

$$\mathscr{A}_{AB} = \mathscr{A}_A + \mathscr{A}_B.$$

Thus the parameter $\mathscr{A}$ *does* show the simple additive behavior characteristic of an extensive property like energy.

In $\mathscr{A}$ we have created a promising index to the behavior of a system. But we do somewhat better by defining a new parameter closely related to $\mathscr{A}$, by writing

$$S \equiv k \ln W, \tag{7}$$

where k is the same universal constant first introduced when we wrote $\beta = 1/k\theta$. The parameter symbolized by S is what we call the *entropy* of the system. Like W and $\mathscr{A}$, S too is increased by any spontaneous change in an isolated system, and S too attains its maximum value when the isolated system achieves its equilibrium condition. Like $\mathscr{A}$, S too represents an extensive property of the system. With all these traits in common, why should S be preferred to $\mathscr{S}$? The reason is simple: incorporation of the constant k in the definition of S serves to place our statistical entropies on a scale completely coincident with the familiar scale of thermodynamic entropies.

To demonstrate this coincidence, we set out again from equation (5), $d \ln W = \beta \, dE$. Multiplying through by the constant k, we draw on the definition embodied in equation (7) to write

$$k \, d \ln W = k\beta \, dE,$$
$$d \, k \ln W = k\beta \, dE,$$
$$dS = k\beta \, dE.$$

Substitution for β, from the relation $\beta = 1/kT$, now yields

$$dS = dE/T.$$

Like equation (5), from which it has been derived, this equation will apply only to constant-volume changes in which energy is transferred solely in the form of heat. Any change in which only $P \, dV$ work is possible will, when conducted under constant-volume conditions, fully satisfy this specification. For if all other forms of work are excluded, and if $P \, dV$ work is then also reduced to zero by the requirement that $dV = 0$, we may indeed be certain that energy can be transferred only in the form of heat.

Consider next the thermodynamic entropies defined by writing $dS_t \equiv q_{rev}/T$, where T symbolizes a temperature shown by thermodynamic analysis to be identical with that read on the ideal-gas scale. An important combination of the first and second laws of classical thermodynamics is

$$dE = T \, dS_t - P \, dV.$$

This equation is restricted to changes in which only $P \, dV$ work is possible. And if we further stipulate constant-volume conditions then, with $dV = 0$, the last equation will be reduced to

$$dS_t = dE/T.$$

The same pair of restrictions applies both to this equation and to the last equation of the preceding paragraph. Hence these equations may be united, to establish that a given energy transfer (dE), made at a given temperature (T), must certainly yield

$$dS = dS_t.$$

And so we have demonstrated that *differences* of entropy measured on the statistical scale defined by equation (7) are, indeed, identical with *differences* of entropy measured on the usual thermodynamic scale.

Since only differences of entropy are thermodynamically significant, we have great latitude in defining a reference zero for the thermodynamic scale of entropies. Indicating by a subscript $_0$ the condition $T \to 0°K$, we ordinarily establish that scale by writing $(S_t)_0 \equiv 0$ for any pure element in a state of internal equilibrium.† And the Nernst heat theorem then justifies the further assertion that $(S_t)_0 = 0$ for any pure compound in a state of internal equilibrium. Hence, for all pure substances

$$(S_t)_0 = 0.$$

The thermodynamic scale so established neglects entropy terms arising from such complications as isotopic mixing, randomness of nuclear spins, etc.— which, to the extent that they contribute alike to the entropies of both the elements and their compounds, are thermodynamically irrelevant. What about the statistical scale of entropies defined by equation (7)? For any pure substance in internal equilibrium at $T \to 0°K$, we again suppose that all the units must fall together in a (hypothetically unique) ground state. Neglecting the complications noted just above, we will then find for the statistical entropy

$$(S)_0 \equiv k \ln W_0 = k \ln N! / \prod \eta_n! = k \ln N!/N! = k \ln 1 = 0.$$

Thus the two scales of entropy can be rendered perfectly coincident in the limit of $T \to 0°K$, and integration of the last equation in the preceding paragraph produces

$$\int_0^S dS = \int_0^{S_t} dS_t$$

$$S = S_t.$$

We are so led to conclude that the statistical scale of entropies, defined in terms of equation (7), is everywhere coincident with the familiar scale of thermodynamic entropies. As obtained with the aid of equations (5) and (6), this conclusion will of course be secure only after completion of two subsequent demonstrations. For, apart from the promised demonstration that $\beta = 1/kT$, we have yet to demonstrate that the applicability of equation (5) extends beyond the assemblies of distinguishable units for which alone it has so far been derived. After examining the statistics of indistinguishable units, in the next chapter, the reader will be invited (on p. 53) to show that they too fall within the scope of equation (5). The above-derived identity of statistical and thermodynamic entropies is thus readily generalizable even to assemblies of indistinguishable units insusceptible to the Boltzmann analysis heretofore employed.

---

† States of internal equilibrium are the implicit presupposition of many thermodynamic statements. The presupposition is here made explicit in order to exclude certain aberrant instances that arise from the extreme slowness with which equilibrium is attained at $T \to 0°K$.

The Partition Function

2

From our earlier statistical analysis, we need carry forward no more than one general conclusion:

In an isolated macroscopic assembly of distinguishable units, the equilibrium configuration is typically that described by the Boltzmann distribution law:

$$\eta_i = \eta_0 e^{-\beta \epsilon_i}. \tag{8}$$

Here η_0 symbolizes the number of units resident in the ground state, and η_i symbolizes the population of the ith quantum state, with energy ϵ_i measured relative to the ground-state energy $\epsilon_0 \equiv 0$. Even after we have learned how to evaluate β, as $1/kT$, equation (8) will offer us but one equation in two unknowns: η_0 and η_i. However we can now easily eliminate η_0 in favor of another quantity more directly accessible to empirical determination.

By definition, the total number (N) of units present in a given assembly must equal the sum of the numbers of units present in each quantum state. Hence

$$N = \eta_0 + \eta_1 + \eta_2 + \cdots = \sum_q \eta_q,$$

where the symbol $\sum_q$ requires a summation over *all* the quantum states characteristic of the units concerned. Equation (8) then permits us to express all the other population numbers in terms of η_0, as

$$N = \eta_0 + \eta_0 e^{-\beta \epsilon_1} + \eta_0 e^{-\beta \epsilon_2} + \cdots = \eta_0 (1 + e^{-\beta \epsilon_1} + e^{-\beta \epsilon_2} + \cdots).$$

With $\epsilon_0 \equiv 0$, we substitute for 1 an equivalent expression in terms of ϵ_0:

$$N = \eta_0 (e^{-\beta \epsilon_0} + e^{-\beta \epsilon_1} + e^{-\beta \epsilon_2} + \cdots) = \eta_0 \sum_q e^{-\beta \epsilon_q}.$$

Consequently

$$\eta_0 = N / \sum_q e^{-\beta \epsilon_q}$$

and, on substitution of this expression for η_0, equation (8) can be recast in the following more generally useful form

$$\eta_i = N \frac{e^{-\beta \epsilon_i}}{\sum\limits_q e^{-\beta \epsilon_q}}. \tag{9}$$

Defining the partition function. The summation shown in equation (9) appears repeatedly in the following discussion, and we do well to assign it a special name and symbol. We call it the *partition function*. This is apt terminology because that function is centrally involved in determining how the units will distribute, or partition, themselves over the possible quantum states. Our symbol for the partition function is z, which derives from the equally apt German term *Zustandssumme*, or sum-over-states. We can then rewrite our last equation in more compact form, as

$$\eta_i = N \frac{e^{-\beta \epsilon_i}}{z},$$

where

$$z \equiv \sum_q e^{-\beta \epsilon_q} = e^{-\beta \epsilon_0} + e^{-\beta \epsilon_1} + e^{-\beta \epsilon_2} + \cdots$$

$$= 1 + e^{-\beta \epsilon_1} + e^{-\beta \epsilon_2} + \cdots$$

$$\text{If } \epsilon_0 \equiv 0$$

The population of the ith quantum state clearly must remain the same *wherever* we choose to set our reference zero of energy, and **problem 13** invites a demonstration that, when all energies are measured consistently from *any one* reference zero, the calculated population η_i does indeed remain unaltered. Ordinarily we choose as reference zero the ground-state energy, for which we then write $\epsilon_0 \equiv 0$; and for this case Gurney has suggested a simple geometric interpretation of the partition function.

In Fig. 11 several bar graphs display the values of the function $e^{-\beta \epsilon_q}$ at the particular ϵ_q's associated with the successive quantum states of two different molecular species. Panel (a) shows the set of bars for some molecule in which those quantum states are rather narrowly separated in energy; panel (b) shows the corresponding bar graph for some other species of molecule in which that energy spacing is much wider. Panels (c) and (d) display the same function for the same respective units when the value of β is three times that for which panels (a) and (b) are drawn—which means for a temperature only one-third that applying in panels (a) and (b). With the ground-state energy set equal to zero, the (axial) line for the ground state is of unit length in every case, and the lengths of all other lines are progressively shorter the higher the energy of the quantum state concerned. Hence, if calculated on the basis of $\epsilon_0 \equiv 0$, the partition function $z(\equiv \sum e^{-\beta \epsilon_q})$ will in each case represent nothing but the *sum* of the lengths of the lines shown in the respective bar graph.

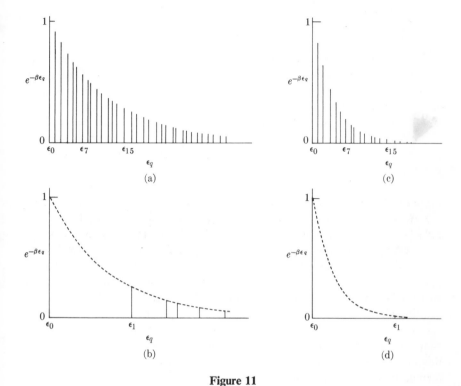

Figure 11

Since $\eta_i = \eta_0 e^{-\beta \epsilon_i}$, the relative lengths of the lines drawn in each panel will reflect the relative populations of the corresponding quantum states. The magnitude of z thus becomes a measure of the extent to which, at any given temperature, the units are distributed over the various quantum states characteristic of their species. In panel (a), for example, the units are spread over many quantum states, and the function $z \gg 1$. With the term $e^{-\beta \epsilon_i}$ representing the contribution of the ith quantum state to the total represented by $\sum_q e^{-\beta \epsilon_q}$, the function $e^{-\beta \epsilon_i}/\sum_q e^{-\beta \epsilon_q}$ obviously expresses the *fraction* of that total contributed by the ith quantum state. Equation (9) then understandably indicates that this fraction of the total number of units (N) will represent the number of units (η_i) resident in the ith quantum state. From panel (a) we see that these populations will be substantial for all quantum states with values of i not too much greater than zero. Panel (b), on the other hand, depicts a situation in which the units are spread only slightly, and here z does not much exceed unity. In this case the populations of all states other than the ground state are comparatively small. And when the same units stand at a lower temperature, as shown in panel (d), essentially all units will fall in the ground state. There is then effectively *no* spread, and we duly find $z = 1$.

The partition function z expresses the sum of a potentially infinite series that converges more rapidly the lower the temperature and the larger the energy spacing between successive quantum states of the units concerned. In evaluating z, we may then terminate the summation as soon as we reach a term $e^{-\beta\epsilon_x} \ll 1$. When, as in panel (d), the energy of even the *first* excited state is so large that $e^{-\beta\epsilon_1} \ll 1$, we may cut off the series after the first (ground-state) term, and so obtain $z = 1$. When, on the other hand, there are many accessible quantum states, as in panel (a), the summation must be carried out over many terms before one comes to a term $e^{-\beta\epsilon_x} \ll 1$. A less tedious evaluation of the partition function may then often be attained by replacement of the summation with an integral that yields a one-step evaluation of the corresponding value of $z \gg 1$.

However it is computed, the magnitude of the partition function z reflects the net accessibility of the various quantum states at the given temperature. And precisely because it does so measure the "spread," the partition function constitutes the essential link between the molecular parameters of the unit and the macroscopic parameters of an assembly of such units. That is, the partition function offers us the possibility of calculating all the thermodynamic parameters of a macroscopic assembly from spectroscopic measurements that establish the energies of the quantum states characteristic of the units constituting that assembly. We can capitalize on this important possibility only after we have mastered in detail the proper formulation and evaluation of partition functions. But before embarking on the somewhat tedious analysis required, we may do well to look at an example that highlights the importance of the partition function. This example demonstrates, for one peculiarly simple case, that knowledge of partition functions represents power to compute the equilibrium constant of an otherwise uncharacterized chemical reaction.

Chemical equilibrium and the partition function. Gas-phase isomerization reactions (e.g., the interconversion of butane and isobutane, or of *cis* and *trans* butene-2) are important molecular equilibria of the type:

$$A\ (g) = B\ (g).$$

Figure 12 offers a highly schematic display of the possible quantum levels of A and B—each set comprising every possible combination of the translational, rotational, vibrational, and electronic quantum states of the respective pure species in its *standard state*. Each energy ϵ_{Ai} or ϵ_{Bi} is measured in the usual way from the respective ground-state energy ϵ_{A0} or ϵ_{B0}. As indicated, the two ground states will ordinarily differ in energy by some quantity we symbolize as $\Delta\epsilon_0^0$. As usual, the superscript ^{0} refers to the standard-state condition, while the subscript $_0$ denotes the temperature of

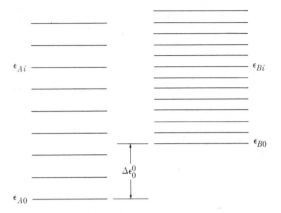

Figure 12

0°K at which all units must fall in their respective ground states. Representing the quantity of energy required to convert one molecule of A in its ground state into one molecule of B in its ground state, $\Delta\epsilon_0^0$ thus expresses the energy change when one molecule of A undergoes the above reaction at 0°K. Concerned as we now are with an equilibrium involving both A and B, we must of course take care to reduce all our energy measurements to a single self-consistent basis. This we can easily do. Setting our reference zero of energy to coincide with the ground state of A, i.e., taking $\epsilon_{A0} \equiv 0$, we need only add the quantity $\Delta\epsilon_0^0$ to each of the B-group energies. That is, for the energy of the ith quantum state of B we will write $(\epsilon_{Bi} + \Delta\epsilon_0^0)$.

Ordinarily the molecules of A have available to them only the A-group of quantum states, and the molecules of B only the B-group of states. However, whenever interconversion of A and B becomes possible (e.g., in the presence of a catalyst), any given molecule will be sometimes A and sometimes B. We can then discuss the situation in terms of a generalized molecule X, which can assume either the A-form or the B-form, and which thus has access to the quantum states of *both* A and B. We have now to consider how the molecules of X will distribute themselves over the A-states and the B-states.

Let us symbolize by η_{A0} the number of X units in the ground state of A. We then call on equation (8) to express as follows the population of the ith quantum state of A:

$$\eta_{Ai} = \eta_{A0}\, e^{-\beta\epsilon_{Ai}}.$$

But the total number of molecules of A is simply the total number (N_A) of X molecules occupying A-states. As the sum of all A-state populations, we write:

$$N_A = \sum_q \eta_{Aq} = \eta_{A0} \sum_q e^{-\beta\epsilon_{Aq}}.$$

Since the summation at the extreme right is nothing but the partition function of pure A in its standard state, evidently

$$N_A = \eta_{A0}\, z_A^0.$$

Proceeding in the same way, for the population (η_{Bi}) of the ith quantum state of B we find:

$$\eta_{Bi} = \eta_{A0}\, e^{-\beta(\epsilon_{Bi}+\Delta\epsilon_0^0)}.$$

The total number (N_B) of molecules of X that appear as B is then

$$N_B = \sum_q \eta_{Bq} = \eta_{A0} \sum_q e^{-\beta(\epsilon_{Bq}+\Delta\epsilon_0^0)} = \eta_{A0}\, e^{-\beta\Delta\epsilon_0^0} \sum_q e^{-\beta\epsilon_{Bq}}.$$

And so

$$N_B = \eta_{A0}\, z_B^0\, e^{-\beta\Delta\epsilon_0^0},$$

where z_B^0 symbolizes the partition function of pure B in its standard state, calculated as usual relative to an energy zero coincident with *its* ground state. However, multiplication by the exponential term suffices to reduce z_B^0 to the energy zero $\epsilon_{A0} \equiv 0$.

With our calculations of both N_A and N_B thus based on the same reference zero of energy, we can divide the equation for N_B by that for N_A, to obtain

$$\frac{N_B}{N_A} = \frac{z_B^0}{z_A^0}\, e^{-\beta\Delta\epsilon_0^0}.$$

But in any given system the ratio of the numbers N_B and N_A must be the same as the ratio of the respective partial pressures, P_B and P_A. Hence, $N_B/N_A = P_B/P_A = K_p$, where K_p is the equilibrium constant expressed in terms of pressures. Again we assume what we have yet to demonstrate: that $\beta = 1/kT$, where the Boltzmann constant $k = R/N$ when R symbolizes the ideal-gas-law constant and N symbolizes Avogadro's number. Making the indicated substitutions, we can then rewrite the last equation as

$$K_p = \frac{z_B^0}{z_A^0}\, e^{-\Delta\epsilon_0^0/kT}. \tag{a}$$

What does this equation teach us about the determinants of the equilibrium condition in the reaction $A(g) = B(g)$? At very low temperatures, when $\Delta\epsilon_0^0 \gg kT$, the equilibrium constant will be minute because $e^{-\Delta\epsilon_0^0/kT} =$ a very small number. This analytical finding is easily interpretable in terms of the distribution law. At very low temperatures, the vast majority of units must fall in the lower-lying A-states, so that equilibrium will be reached with practically the entire contents of the system in form A. Far different is the situation at temperatures high enough that $\Delta\epsilon_0^0 \ll kT$. In these circumstances the exponential factor has little weight, because now $e^{-\Delta\epsilon_0^0/kT} \simeq e^{-0} = 1$. And so at high temperature it is the ratio z_B^0/z_A^0 that must become controlling.

If $z_B^0 > z_A^0$, the formation of B will be favored at high temperature even though (as an endothermic reaction) the formation of B is disfavored at low temperature.

Compare with this the perspective afforded by classical thermodynamics, from which we draw the two familiar equations:

$$\Delta G_T^0 = \Delta H_T^0 - T \, \Delta S_T^0 \quad \text{and} \quad \Delta G_T^0 = -RT \ln K_p.$$

Combination of these yields

$$\ln K_p = -\frac{\Delta H_T^0}{RT} + \frac{\Delta S_T^0}{R}.$$

At sufficiently low temperatures, the first term on the right must be dominant, and the endothermic conversion of A into B is disfavored. However, with rise of temperature, this first term (with T in the denominator) will become progressively less important. Ultimately the second term must become dominant, and the reaction that proceeds with positive ΔS_T^0 will then proceed with an equilibrium constant $K_p > 1$. Thus the thermodynamic analysis tallies with the statistical analysis, in that both conceive K_p to be a function of two terms: a ΔH_T^0 (or $\Delta \epsilon_0^0$) term dominant at low temperatures, and a second term dominant at high temperatures. But, though very simple and completely general, the thermodynamic analysis fails to yield something we gain from the statistical analysis: a profound insight into the *why* of the predicted events. For one thing, in terms of the lower placement of the A-states, we see clearly why the endothermic reaction is disfavored at low temperatures. And, even more important, in the statistical analysis the sign and magnitude of the thermodynamically opaque ΔS_T^0 term become readily interpretable. How?

Returning again to equation (a), we observe that $\Delta \epsilon_0^0/kT = N \, \Delta \epsilon_0^0/NkT = \Delta E_0^0/RT$. Where $\Delta \epsilon_0^0$ symbolized the energy change *per molecule* undergoing the indicated reaction, ΔE_0^0 represents the energy change *per mole* for the indicated reaction at $0°K$. Equation (a) may then be rewritten, in logarithmic form, as

$$\ln K_p = -\frac{\Delta E_0^0}{RT} + \frac{R \ln (z_B^0/z_A^0)}{R}.$$

Compare this with the last thermodynamic equation. Insofar as ΔE_0^0 is equivalent to ΔH_T^0 for our simple gas-phase reaction, to the same (excellent) degree of approximation $\Delta S_T^0 = R \ln (z_B^0/z_A^0)$. If at high temperature $z_B^0 > z_A^0$, then the reaction will necessarily proceed with $\Delta S_T^0 > 0$. For once $kT \gg \Delta \epsilon_0^0$, the units will spread themselves fairly evenly over all accessible quantum states, and the proportion of B to A in the equilibrium mixture will be determined primarily by the relative numbers of accessible B-states and A-states. Even though the (endothermic) formation of B is disfavored at low temperatures, a large yield of B becomes entirely possible if, at high temperatures, the accessible B-states far outnumber the accessible A-states. (Conversely, if the A-states were more numerous than the B-states, we would be unable to obtain even a 50% yield of B at *any* temperature.) Thus, for the equilibrium condition in the model reaction $A(g) = B(g)$, the statistical analysis presents an account that is not only comprehensive but, in addition, thoroughly comprehensible. We see then ample prospect of reward in the now impending analysis of the proper formulation and evaluation of partition functions.

FORMULATION OF PARTITION FUNCTIONS

Our statistical analysis has throughout taken for granted the possibility of assigning a unique "private" energy to each unit of an assembly. Clearly this assumption fails in highly interactive systems, for no energies will be *individually* assignable to units that interact strongly with their neighbors. Thus led to stipulate effective independence of constituent units, we *cannot* however insist on their complete independence. For as noted earlier (on p. 24) *some* interaction between units is prerequisite to the energy exchanges required for production of the equilibrium configurations that remain our primary concern. And so we are prompted to restrict our consideration to assemblies of "weakly coupled" units: units that are effectively independent but sufficiently interactive to ensure eventual attainment of true equilibrium states. Fortunately this compromise requirement of weak coupling is reasonably well satisfied in many actual systems, e.g., a real gas at moderate pressure.

Multiple degrees of freedom. In formulating the partition function, we must acknowledge the fact that polyatomic gas molecules may store energy in diverse ways: in their translational, rotational, and vibrational motions, and also in the form of electronic excitation. How can we formulate a partition function that duly allows for these different "degrees of freedom"? We need only call again on an assumption of weak coupling. Already stipulated for the units of an assembly, a condition of weak coupling is now further stipulated to characterize the relations among the various degrees of freedom possessed by any one unit in that assembly. Though the possibility of energy exchanges among these degrees of freedom is thus preserved, effective independence of the various degrees of freedom is still ensured. The fact that a given unit is in some particular translational state, say, then has no bearing on its rotational state, and conversely. To the extent that this plausible but essentially classical conception is valid, the partition function for each such weakly coupled degree of freedom can be formulated as though it were the *only* degree of freedom possessed by the unit. The sole question then remaining is how these individual partition functions are to be combined in constructing the total partition function for the molecular unit in question.

For a unit with many degrees of freedom, we can express the total partition function as $z_{tot} \equiv \sum_q e^{-\beta \epsilon_q}$, where the usual summation over every possible quantum state here entails summation over every possible *combination* of the unit's translational, rotational, vibrational, and electronic quantum states. Permitting us to regard each degree of freedom as essentially unperturbed by the others, the assumption of weak coupling allows us to express the total energy of the unit as the sum of energies individually associated with each of the several degrees of freedom. Thus

$$z_{tot} = \sum_q e^{-\beta(\epsilon_{tr} + \epsilon_{rot} + \epsilon_{vib} + \epsilon_{elec})}, \tag{b}$$

where the summation still extends over all possible quantum states of the unit. What this means can most readily be seen for a unit with, say, only rotational and vibrational degrees of freedom. Let the energies associated with the divers rotational and vibrational states be denoted by subscripts $r0, r1, r2, \ldots$, and $v0, v1, v2, \ldots$ respectively indicating the corresponding rotational and vibrational quantum numbers. For the overall partition function (z_{comb}) we can then write:

$$
\begin{aligned}
z_{comb} &= e^{-\beta(\epsilon_{r0} + \epsilon_{v0})} + e^{-\beta(\epsilon_{r0} + \epsilon_{v1})} + e^{-\beta(\epsilon_{r0} + \epsilon_{v2})} + \cdots \\
&\quad + e^{-\beta(\epsilon_{r1} + \epsilon_{v0})} + e^{-\beta(\epsilon_{r1} + \epsilon_{v1})} + \cdots \\
&\quad + e^{-\beta(\epsilon_{r2} + \epsilon_{v0})} + \cdots + \cdots \\
&= e^{-\beta\epsilon_{r0}} \cdot e^{-\beta\epsilon_{v0}} + e^{-\beta\epsilon_{r0}} \cdot e^{-\beta\epsilon_{v1}} + \cdots \\
&\quad + e^{-\beta\epsilon_{r1}} \cdot e^{-\beta\epsilon_{v0}} + \cdots + \cdots \\
&= (e^{-\beta\epsilon_{r0}} + e^{-\beta\epsilon_{r1}} + \cdots) \times (e^{-\beta\epsilon_{v0}} + e^{-\beta\epsilon_{v1}} + \cdots) \\
&= \sum_{rq} e^{-\beta\epsilon_{rq}} \times \sum_{vq} e^{-\beta\epsilon_{vq}},
\end{aligned}
$$

where the summations extend over all possible values of the rotational quantum number (rq) and the vibrational quantum number (vq) respectively. But these summations are simply the partition functions for independent rotational and vibrational degrees of freedom, so that

$$z_{comb} = z_{rot} \cdot z_{vib}.$$

And returning then to the general case of equation (b), we see that the same analysis must there yield

$$z_{tot} = z_{trans} \cdot z_{rot} \cdot z_{vib} \cdot z_{elec}. \tag{10}$$

For a species of unit with multiple degrees of freedom, we can thus easily formulate the overall partition function in terms of individual partition functions severally evaluated for each single degree of freedom. One may now think to object that the indicated degrees of freedom aren't really singular. Even a diatomic molecule has two rotational degrees of freedom;

and nonlinear polyatomic molecules possess three rotational degrees of freedom, as well as multiple vibrational degrees of freedom. How shall we cope with all these? Very simply indeed: we need only push ahead with exactly the kind of analysis already begun. As an example, consider an ordinary gas particle possessing *not* one but three translational degrees of freedom—representing components of motion parallel to the x-, y-, and z-axes. Each of these degrees of freedom can be assumed independent of the others, i.e., the magnitude of the velocity component parallel to the x-axis does not at all predetermine the y- or z-components, and conversely. We can then separately formulate a partition function (z_{trx}) representing the possible quantum states for motion paralleling the x-axis, as well as the corresponding functions (z_{try} and z_{trz}) for motions parallel to the y- and z-axes respectively. Embodying every possible combination of all the terms appearing in the separate partition functions for the x-, y-, and z-directions, the complete translational partition function (z_{trans}) can then be written:

$$z_{trans} = z_{trx} \cdot z_{try} \cdot z_{trz}.$$

Provided that all the degrees of freedom are adequately independent of one another, we can in this fashion progressively work our way back to the component individuals for each of which a separate partition function is readily formulated. And conversely, by multiplying these component partition functions, we easily arrive at the total partition function for a unit with multiple degrees of freedom.

Degeneracy. The Boltzmann distribution law, as we have derived it, refers to the population of each distinct quantum state characteristic of the units under consideration. In certain instances the complete set of quantum states may include some which, though otherwise distinct, are very nearly or exactly the same in energy. Thus, irrespective of the orientation in space of the axis of rotation, the energy associated with a given molecular rotor is ordinarily determined solely by the value of an integral quantum number (J). With a classical macroscopic rotor the orientation angle (relative to an external electric or magnetic field) is continuously variable, but in quantum physics the angle of orientation of a molecular rotor is restricted to just ($2J + 1$) possibilities corresponding to each integral value of the quantum number J. Each such orientation represents one distinct quantum state of the rotor even though, in the absence of a field, the energy associated with the rotation remains unaffected by the angle—just as in the classical case. Consequently, the same energy, dependent only on the magnitude of J, will characterize all members of the group of ($2J + 1$) rotational quantum states. How can we best formulate the partition function in these circumstances?

As a specific example, consider that the same energy (ϵ_1) is associated with each of the three distinct rotational states for which $J = 1$. Denoting each of the three by an appropriate number of primes, we use equation (9) to establish the population of each of these quantum states:

$$\eta_1' = N \frac{e^{-\beta\epsilon_1}}{\sum_q e^{-\beta\epsilon_q}}, \qquad \eta_1'' = N \frac{e^{-\beta\epsilon_1}}{\sum_q e^{-\beta\epsilon_q}}, \qquad \eta_1''' = N \frac{e^{-\beta\epsilon_1}}{\sum_q e^{-\beta\epsilon_q}}.$$

In every case the right-hand side of the equation is obviously identical. Symbolizing by η_1^* the *total* number of units in *all three* quantum states, we then easily conclude that

$$\eta_1^* = \eta_1' + \eta_1'' + \eta_1''' = N \frac{3e^{-\beta\epsilon_1}}{\sum_q e^{-\beta\epsilon_q}}.$$

The number 3 has appeared simply because, in calculating η_1^*, we have lumped together three distinct but energetically equivalent quantum states. When three such quantum states are grouped together in one composite energy level, we speak of that energy level as "threefold degenerate." When ω distinct quantum states are equivalent in energy, we may collect them in a single energy level we then speak of as "ω-fold degenerate." (An energy level comprising only one quantum state will of course be assigned a degeneracy $\omega = 1$.) For *any* energy level i, with energy ϵ_i and degeneracy ω_i, the same reasoning that yielded the last equation will permit us to write for the general case

$$\eta_i^* = N \frac{\omega_i e^{-\beta\epsilon_i}}{\sum_q e^{-\beta\epsilon_q}},$$

where η_i^* symbolizes the total population of the single energy level that comprises ω_i quantum states which, though absolutely distinct in principle, are quite undifferentiated with respect to energy.

For any given assembly, the total number (N) of units in all quantum states must of course be the same as the total number of units in all energy levels. Summing over all the *energy* levels, we thus find

$$N = \eta_1^* + \eta_2^* + \eta_3^* + \cdots = \sum_y \eta_y^*$$

$$= N \frac{\omega_0 e^{-\beta\epsilon_0}}{\sum_q e^{-\beta\epsilon_q}} + N \frac{\omega_1 e^{-\beta\epsilon_1}}{\sum_q e^{-\beta\epsilon_q}} + \cdots = N \frac{\sum_y \omega_y e^{-\beta\epsilon_y}}{\sum_q e^{-\beta\epsilon_q}}.$$

Just as the index q indicates a summation over all quantum states characterized by energies ϵ_q, the index y here indicates a summation over all energy levels characterized by energies ϵ_y and degeneracies ω_y. The last equation demonstrates the absolute equality of the two summations. Dropping the indices amply suggested by the corresponding subscripts within the summations, we see that the same partition function (z) can be obtained by either of two fully equivalent accounting procedures:

$$\underset{\substack{\text{Reckoned for individual} \\ \text{quantum states}}}{\sum e^{-\beta\epsilon_q}} = z = \underset{\substack{\text{Reckoned for degenerate} \\ \text{energy levels}}}{\sum \omega_y e^{-\beta\epsilon_y}}. \tag{11}$$

We may then evaluate z by summing *either* over all the individual quantum states, *or* over all the energy levels comprising groups of energetically equivalent quantum states. If we draw up our account in terms of energy levels, any particular ω_i-degenerate level with energy ϵ_i will be represented in the partition function ($= \sum \omega_y e^{-\beta \epsilon_y}$) by the *product* $\omega_i \times e^{-\beta \epsilon_i}$. If instead we draw up our account in terms of quantum states, the ω_i distinct quantum states with energy ϵ_i will be represented in the partition function ($= \sum e^{-\beta \epsilon_q}$) by the *sum* of ω_i identical $e^{-\beta \epsilon_i}$ terms. Quite obviously, the product and the sum have exactly the *same* magnitude. Thus we see clearly why—if we are consistent in our application of whatever accounting procedure we have chosen for the job in hand—the value of z must remain wholly independent of our choice. May one then conclude (with disappointment) that this entire exploration of degeneracy has eventuated in nothing more than an occasionally convenient but nonessential alternative method for evaluating the partition function z? Certainly not! For we now demonstrate that the concept of degeneracy will permit us to resolve a hitherto unmentioned but absolutely fundamental matter of principle.

Indistinguishable units. Boltzmann statistics is characteristically the statistics of *distinguishable* units. To be sure, we drew the Boltzmann distribution law from an analysis of assemblies of units stipulated to be identical, and hence not *intrinsically* distinguishable. However, no difficulty arises as long as the units are *localized,* as are harmonic oscillators locked in a crystal lattice. For then, even though the units are themselves indistinguishable, we can in principle identify them in terms of their permanent occupation of different positions in the crystal. Thus we can properly distinguish the *oscillator occupying position* x from the *oscillator occupying position* y. But no such distinction will be possible when (as in our analysis of isomerization equilibria) we are concerned with assemblies of gas molecules in free translational motion. No longer localized, identical units of this sort are no longer distinguishable even in principle, and the whole application of a Boltzmann statistics presuming such distinguishability is then called in question. In these circumstances we cannot even take for granted the applicability of the Boltzmann distribution law on which we have based our formulation of the partition function.

    Can the Boltzmann distribution law be applied to assemblies of ordinary gas molecules in free translational motion? In a rigorous sense, the answer is "No!" But in a practical sense, the answer is "Yes!" The feasibility of this extension of the Boltzmann law arises from the saving circumstance that, ordinarily, the number of accessible translational quantum states vastly exceeds the number of units distributed over those states. We defer to p. 76 the rather tedious demonstration of this "dilute gas" condition, the consequence of which is immediately obvious. That is, when the number of accessible quantum states vastly exceeds the number of units present, the great majority of states must remain untenanted, a comparative few will be singly tenanted, and effectively *none* will have populations in excess of 1. We now demonstrate that in these circumstances the Boltzmann law can be extended to assemblies of indistinguishable units in free translational motion.

At first sight, what we have called the "saving circumstance" may appear anything but that. For one thing, we obtained the Boltzmann distribution law from an analysis involving the displacement of $(p + q)$ units, where p and q represent integers stipulated to be "small" in comparison with the population numbers. Such an analysis appears impossible when all the population numbers are either 0 or 1, relative to which *no* integer values of p and q can be regarded as "small." An even more fundamental objection is that, given such population numbers, the whole concept of a predominant configuration may seem to become meaningless, since the equation $W = N!/\prod \eta_n!$ would then yield $W = N!$ for *all* configurations of an assembly of distinguishable units. Fortunately, one and the same simple expedient serves to resolve both these difficulties.

Rather than considering the populations of individual *quantum states*, let us instead consider the populations of *energy levels* comprising huge numbers of quantum states effectively equivalent in energy. The feasibility of this expedient is fully ensured by a corollary of the "dilute gas" condition. For if the number of translational quantum states vastly exceeds the number of units—a number no less than 6×10^{23} for a mole of gas molecules—it follows necessarily that these quantum states must be *very* narrowly separated in energy. Enormous numbers of quantum states nearly identical in energy can then be grouped together in a relatively small set of energy levels with suitably large *total* populations. Given such populations, we *can* properly ask what distribution of the units over the energy levels will represent the predominant configuration in which the number of distinct microstates (W) achieves its maximum possible value. And that question may be answered, just as before, by way of an analysis of the displacement of $(p + q)$ units, where the integers p and q *can* now be regarded as "small" relative to the total population numbers.

In how many different ways can η_i^* indistinguishable units be distributed over the ω_i energetically equivalent but distinct quantum states comprised within an ith energy level? To obtain an answer, we need only solve a classical problem: in how many different ways can η_i^* identical balls be distributed over ω_i identifiable boxes? Imagine the boxes placed close together, in a linear array. In Fig. 13 the line AZ symbolizes the extent of this array, and the points A and Z indicate the locations of the outermost walls of the first and last boxes respectively. If the array is to comprise a total of ω_i boxes, just $(\omega_i - 1)$ partitions must appear between the end walls at A and Z. Thus the second panel of the figure shows how four boxes are created by insertion of the three (double-wall) dividers symbolized as $\prod$. Each such box is unequivocally distinguishable in terms of its placement in

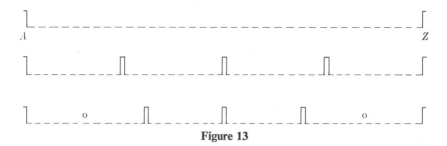

Figure 13

the sequence that extends from A to Z. For example, of the ten different ways in which two identical balls (symbolized as o) can be distributed over four identifiable boxes, the third panel of the figure displays the distribution in which box 1 contains one ball, boxes 2 and 3 contain none, and box 4 again contains one ball.

To determine in how many different ways η_i^* identical balls can be distributed over the ω_i identifiable boxes, we need now only determine the number of distinguishably different sequences in which the η_i^* o-symbols and the $(\omega_i - 1) \sqcap$ -symbols can be arranged along the line AZ. For the first term in the sequence we may choose any one of the available $(\eta_i^* + \omega_i - 1)$ symbols; for the next term, any one of the $(\eta_i^* + \omega_i - 2)$ symbols then remaining; and so on until only one symbol remains for entry as the last term of the sequence. Clearly the total number of possible sequences is $(\eta_i^* + \omega_i - 1)!$, and these sequences must embrace *every* possible distribution of the balls over the boxes. On the other hand, given the postulated identity of the η_i^* o-symbols and of the $(\omega_i - 1) \sqcap$ -symbols, many of the possible sequences will represent the *same* distribution. Thus all the $\eta_i^*!$ sequences arising solely from permutations of the identical o-symbols, and all the $(\omega_i - 1)!$ sequences arising solely from permutations of the identical $\sqcap$ -symbols must yield one and the same apparent distribution. As the net number (W_i) of recognizably *different* sequences, we then find

$$W_i = \frac{(\eta_i^* + \omega_i - 1)!}{\eta_i^*! \, (\omega_i - 1)!} \cdot$$

One may wish to confirm that this function does indeed yield for W_i the value of 10 easily shown to be the number of different ways in which two identical balls can be distributed over four distinctive boxes.

Yielding the number of ways in which η_i^* identical units can be distributed over ω_i energetically equivalent but distinct quantum states, the last equation represents the fundamental expression of a (Bose-Einstein) quantum statistics applying to indistinguishable units for which multiple occupancy of the same quantum state is permissible. **Problem 14** indicates how a rather different equation is obtained with the (Fermi-Dirac) quantum statistics applying to indistinguishable units for which such multiple occu-

pancy is forbidden. But both these systems of statistics yield the same result—
a result ultimately compatible with Boltzmann statistics—whenever conformity with the "dilute gas" condition ensures that $\omega_i \gg \eta_i^*$.

To obtain that result, we expand the factorial in the numerator of the
last equation, as follows:

$$W_i = \frac{(\eta_i^* + \omega_i - 1)(\eta_i^* - 1 + \omega_i - 1) \cdots (1 + \omega_i - 1)(\omega_i - 1)!}{\eta_i^*! \, (\omega_i - 1)!}$$

$$= \frac{(\eta_i^* + \omega_i - 1)(\eta_i^* - 1 + \omega_i - 1) \cdots (1 + \omega_i - 1)}{\eta_i^*!}.$$

With $\omega_i \gg \eta_i^*$, the numerator represents a product of η_i^* terms all with
magnitude of the order of ω_i. An excellent approximation is then simply

$$W_i \simeq \frac{(\omega_i)^{\eta_i^*}}{\eta_i^*!}.$$

Of course, W_i represents *only* the number of microstates possible when η_i^*
indistinguishable units are spread thinly over the ω_i quantum states comprised in the ith energy level. But since any microstate of the ith energy
level may combine with any microstate of each other energy level—to form
one distinct microstate of the assembly—we may be sure that the total
number (W) of possible microstates will be given as

$$W = \prod W_y = \frac{\prod (\omega_y)^{\eta_y^*}}{\prod \eta_y^*!}, \tag{12}$$

where the products extend over all the y energy levels in terms of which we
have conducted our analysis.

While the equation $W = N!/\prod \eta_n!$ applied to an assembly of N identical
units distinguished by their localization, the last equation is the corresponding expression for N indistinguishable units (for example, gas molecules) in
free translational motion. And just as before we sought a description of
the (predominant) configuration for which the equation $W = N!/\prod \eta_n!$ yields
a maximum value of W, so now we seek a description of the (predominant)
configuration for which the last equation yields a maximum value of W. In
both cases the analysis is the same in all essentials.

We select any three energy levels (l, m, and n) the energy spacing of
which is expressible as

$$\frac{\epsilon_n - \epsilon_m}{\epsilon_m - \epsilon_l} = \frac{p}{q},$$

where p and q are integers small in comparison with the respective population numbers η_l^*, η_m^*, and η_n^*. Starting now from equation (12), we derive
an expression for the change in W consequent to the usual minimum dis-

placement of $(p + q)$ units. Noting that for the predominant configuration this displacement must produce effectively *no* change in W, **problem 15** invites a demonstration that the same style of analysis used before will here yield, as a description of the predominant configuration,

$$\left[\frac{\eta_l^*/\omega_l}{\eta_m^*/\omega_m}\right]^p = \left[\frac{\eta_m^*/\omega_m}{\eta_n^*/\omega_n}\right]^q.$$

Remarking the resemblance of this expression to equation (c) on p. 18, we are encouraged to proceed still further along the same line as before. Rewriting the last expression in logarithmic form, we substitute for the ratio p/q the equivalent ratio of energy spacings. We then soon arrive at a hauntingly familiar equation. As a description of the predominant configuration of an assembly of indistinguishable units in free translational motion, we find

$$\frac{\eta_i^*}{\omega_i} = \frac{\eta_0^*}{\omega_0} e^{-\beta\epsilon_i} \qquad \text{(c)}$$

where ϵ_i symbolizes the energy of the ith energy level, measured relative to the reference zero established by writing for the ground level $\epsilon_0 \equiv 0$.

Compare this last equation with equation (8), $\eta_i = \eta_0 e^{-\beta\epsilon_i}$. With η_i^* units in an ω_i-fold degenerate energy level, the quotient η_i^*/ω_i represents the *average* population per individual quantum state, and is thus a quantity properly correspondent to η_i. And exactly the same may be said of the quotient η_0^*/ω_0 and the ground-state population η_0. For indistinguishable units in free translational motion, we have thus found in equation (c) a description of the predominant configuration that is fully equivalent to the Boltzmann distribution law originally derived for identical units rendered distinguishable by their localization. We may then hope to obtain for unlocalized units a partition-function expression wholly equivalent to that earlier obtained for units distinguished by their localization.

Summing over all energy levels, to obtain the total number of units (N) present in the given assembly, we go on to substitute for the population numbers the expressions supplied by equation (c):

$$N = \eta_0^* + \eta_1^* + \eta_2^* + \cdots = \eta_0^* + \frac{\eta_0^*}{\omega_0}\omega_1 e^{-\beta\epsilon_1} + \frac{\eta_0^*}{\omega_0}\omega_2 e^{-\beta\epsilon_2} + \cdots$$

With $\epsilon_0 \equiv 0$, we substitute for η_0^* an equivalent expression in terms of ϵ_0

$$N = \frac{\eta_0^*}{\omega_0}\omega_0 e^{-\beta\epsilon_0} + \frac{\eta_0^*}{\omega_0}\omega_1 e^{-\beta\epsilon_1} + \frac{\eta_0^*}{\omega_0}\omega_2 e^{-\beta\epsilon_2} + \cdots$$

$$= \frac{\eta_0^*}{\omega_0}\sum \omega_y e^{-\beta\epsilon_y} = \frac{\eta_0^*}{\omega_0} z,$$

where z is simply a partition function that equation (11) shows to be identical with our original partition function $z \equiv \sum e^{-\beta\epsilon_q}$. Therefore

$$\frac{\eta_0^*}{\omega_0} = \frac{N}{z}.$$

And, on substituting in equation (c), we recover an expression easily seen to be entirely equivalent to equation (9):

$$\frac{\eta_i^*}{\omega_i} = \frac{N}{z} e^{-\beta\epsilon_i}. \tag{13}$$

To this extent, we can then treat gas molecules in free translational motion just as we formerly treated identical units rendered distinguishable by their localization. And as earlier promised, **problem 16(a)** now invites a demonstration that equation (5) applies to assemblies of gas molecules even as it was formerly shown to apply to assemblies of localized units. But these two kinds of assembly show also one highly significant difference to which we now address ourselves.

Multiple units. Though in our discussion of the partition function we have spoken repeatedly of an N-unit assembly, the actual magnitude of z is entirely independent of the number of units present. That is, when we write $z \equiv \sum e^{-\beta\epsilon_q} = \sum \omega_y e^{-\beta\epsilon_y}$, with the summations extending respectively over all the quantum states or energy levels characteristic of single units, z clearly symbolizes the partition function *per unit*. Let us now define a partition function (Z) for the assembly as a whole. Proceeding in an analogous fashion, we write $Z \equiv \sum e^{-\beta E_q} = \sum \omega_y e^{-\beta E_y}$, with the summations now extending respectively over all the quantum states or energy levels of the *assembly as a whole*. How can we express Z in terms of z, which we already know how to formulate?

Let us first examine the case of localized units. With reference to the schematic diagram shown below, consider the following two situations: (I) the unit occupying site a is in vibrational state 1, while the unit occupying site b is in vibrational state 2; and, with all other units remaining in their original states, (II) the unit occupying site a is in vibrational state 2, while the unit occupying site b is in vibrational state 1.

$$
\begin{array}{|ccc|}
\hline
a & b & c \\
\cdot & \cdot & \cdot \\
& & \\
\cdot & \cdot & \cdot \\
d & e & f \\
\hline
\end{array}
$$

These are in principle two quite distinct states of the assembly and, given this element of distinctness, one easily sees what must be the relation of Z to z.

For simplicity, let us begin with an assembly of just two units distinguished by localization in positions a and b respectively. With units that are only weakly coupled, the total energy (E_q) of any quantum state of the assembly must be the sum of the individual energies characteristic of the quantum states of the units a and b. Hence

$$Z_{ab} = \sum e^{-\beta E_q} = \sum e^{-\beta(\epsilon_a + \epsilon_b)}.$$

Extending over all possible quantum states of the assembly, the summation must incorporate every possible combination of the quantum states of the individual units.

Using subscripts to denote the quantum states of those units, we can now proceed exactly as we did on p. 44. That is

$$Z_{ab} = e^{-\beta(\epsilon_{a0} + \epsilon_{b0})} + e^{-\beta(\epsilon_{a0} + \epsilon_{b1})} + e^{-\beta(\epsilon_{a0} + \epsilon_{b2})} + \cdots$$

$$+ e^{-\beta(\epsilon_{a1} + \epsilon_{b0})} + e^{-\beta(\epsilon_{a1} + \epsilon_{b1})} + \cdots$$

$$+ e^{-\beta(\epsilon_{a2} + \epsilon_{b0})} + \cdots$$

$$= e^{-\beta\epsilon_{a0}} \cdot e^{-\beta\epsilon_{b0}} + e^{-\beta\epsilon_{a0}} \cdot e^{-\beta\epsilon_{b1}} + \cdots$$

$$+ e^{-\beta\epsilon_{a1}} \cdot e^{-\beta\epsilon_{b0}} + \cdots$$

$$= (e^{-\beta\epsilon_{a0}} + e^{-\beta\epsilon_{a1}} + \cdots) \times (e^{-\beta\epsilon_{b0}} + e^{-\beta\epsilon_{b1}} + \cdots)$$

$$= \sum e^{-\beta\epsilon_{aq}} \times \sum e^{-\beta\epsilon_{bq}},$$

where the summations extend over all possible quantum states of the individual units a and b. But these summations are simply the partition functions of the two individual units, so that $Z_{ab} = z_a \cdot z_b$. Given that a and b are identical, the numerical values of z_a and z_b must be the same—simply because the matching terms in both series will be equal. Hence we may drop the subscripts and, for this two-unit assembly, we write: $Z = z^2$. And a perfectly straightforward extension of the same line of argument yields a corresponding expression for the partition function (Z_{dist}) of an assembly of N identical but distinguishable units:

$$Z_{\text{dist}} = z^N. \tag{d}$$

How will matters differ when identical units are *not* distinguishable? For an assembly of gas molecules in free translational motion, consider the following two situations: (I) the unit referred to as a is in translational state 1, while the unit referred to as b is in translational state 2; and, with all other units remaining in their original states, (II) the unit referred to as a is in translational state 2, while the unit referred to as b is in translational state 1. Do situations (I) and (II) constitute distinct states of the assembly? With an assembly of identical harmonic oscillators locked in a crystal lattice, unit a ("upper-left-corner oscillator") is in principle differentiated from unit *b.* The letter indices then signify genuine distinctions, and situations (I) and (II) represent *different* states of the assembly. But in an assembly of identical gas molecules every spatial juxtaposition of the units is transient, and immediately succeeded by others. Since nothing now differentiates one identical unit from another, the letter indices become completely meaningless, and situations (I) and (II) must represent the *same* state of the assembly. What will then be the relation of Z to z?

The tabulation shown below displays the 3! = 6 distinct ways in which three different quantum states can be assigned to three localized units meaningfully distinguished by letter indices. As units that cannot thus be distinguished, consider three gas molecules moving freely in a container. Instead of six distinct *combinations* (of letters and numbers) there now remain only six *permutations* (of numbers) all of which represent *one and the same quantum state* of the assembly. For all that can be said of any such permutation must be said alike of every such permutation: "one unit occupies state 1, one unit occupies state 2, and one unit occupies state 3." Were we to consider any other group of three different quantum states (say 4, 5, and 6), we would again note six ways in which they can be assigned to three distinguishable units, but only one way in which they can be assigned to three units among which there is no meaningful physical distinction.

a	b	c
1	2	3
1	3	2
3	1	2
2	1	3
2	3	1
3	2	1

The partition function Z is found by summing over every distinct quantum state of the assembly. For an assembly of three distinguishable units, the summation yielding Z_{dist} will contain groups of six numerically equal terms—one for each of the six distinct states of the assembly that arise from different assignments to the units of, say, quantum states 1-2-3. But in the summation yielding Z_{trans}, for an assembly of three identical units in free translational motion, each six-term group is reduced to a single term—representing the solitary state of the assembly that arises from every possible assignment to the units of, say, quantum states 1-2-3. Like the number of distinct states of the assembly, the numerical value of the partition function must then also be reduced by the factor 1/3!. That is

$$Z_{\text{trans}} = \frac{1}{3!} Z_{\text{dist}}.$$

Combination of this finding with equation (d) then yields

$$Z_{trans} = \frac{1}{3!} (z)^3,$$

where z symbolizes the translational partition function *per unit*, and $1/3!$ is the factor that allows for the nondistinguishability of the three units in the assembly.

Can we generalize the above argument to assemblies of N identical units in free translational motion? Indeed we can! For an assembly of N distinguishable units, no less than $N!$ distinct states of the assembly arise from redistributions of the units over any set of N different quantum states. But for an assembly of N indistinguishable units, all these $N!$ states collapse into one and the same quantum state of the assembly. The partition function Z, calculated by summing over all distinct states of the assembly, must then undergo a similar shrinkage when we pass from the assembly of N distinguishable units to the assembly of N identical units in free translational motion. Hence:

$$Z_{trans} = \frac{1}{N!} (z)^N. \tag{14}$$

One should be quite clear that the factor $(1/N!)$ will be appropriate only if essentially all units occupy *different* quantum states. For if, of N distinguishable units, η_1 fall in quantum state 1, η_2 in quantum state 2, and so on, exchanges of the units would produce not $N!$ but only $N!/\prod \eta_n!$ distinct states of the assembly. Thus the last equation will apply only when the number of accessible translational quantum states so exceeds the number of units present that effectively no two units will fall in the same quantum state. Below about 3°K the accessibility of the translational quantum states is so far reduced that this condition may no longer prevail. In this range of temperature, helium is the only substance that can still be handled as a gas, and the peculiar behavior displayed by helium at very low temperature is in part attributable to the fact that the number of accessible quantum states does *not* then vastly exceed the number of units present. In such circumstances one is forced to abandon Boltzmann statistics in favor of Bose-Einstein or Fermi-Dirac statistics. Fortunately, however, this complication is quite irrelevant to our concern with atomic and molecular assemblies at temperatures well removed from 0°K. For the "dilute gas" condition is then fully satisfied, and we need have no hestitation in applying equation (14).

Compare equation (14) with equation (d): $Z = z^N$. Between these two expressions for Z as $f(z)$, our choice is determined by whether or not the units concerned may be regarded as distinguishable. In deriving equation (d), we supposed that identical units are rendered distinguishable by localization; but distinguishability may develop in quite a different way. Consider an assembly of identical units which, while in free translational motion, have also some internal degree(s) of freedom, for example, vibration. In evaluating the partition function Z_{trans}, we must of course use equation (14). But on turning then to the vibrational degree of freedom, we are no longer discussing undifferentiated units. For, with regard to any degree(s) of freedom *apart from* translation, each otherwise identical unit will be distinguished by its occupation of a different translational quantum state. In evaluating the vibrational partition function of a two-unit assembly, for example, the following must certainly be counted as two wholly distinct states of the assembly: (I) the unit occupying translational state a is in vibrational state 1, while the unit occupying translational state b is in vibrational state 2; and (II) the unit occupying translational state a is in vibrational state 2, while the unit occupying translational state b is in vibrational state 1. In this context the differentiation is in principle as well established by occupation of distinctive translational quantum states as by occupation of distinctive spatial locations. *Whether or not* units are localized, they are then to be counted as distinguishable when it becomes a matter of constructing Z for any "internal" degree of freedom like rotation, vibration, or electronic excitation. For such degrees of freedom, therefore, the appropriate relation is *always* that given by equation (d), which we now rewrite as

$$Z_{\text{int}} = z^N. \tag{15}$$

Having learned how to formulate Z for each individual degree of freedom, we turn lastly to formulation of the overall partition function for an assembly of units having multiple degrees of freedom. This overall partition function we define by writing $Z_{\text{tot}} \equiv \sum e^{-\beta E_{\text{tot}}}$ where, for weakly coupled degrees of freedom, the assembly's total energy (E_{tot}) will be expressible as the sum of energies severally associated with the different degrees of freedom. That is, $E_{\text{tot}} = E_{\text{trans}} + E_{\text{rot}} + E_{\text{vib}} + E_{\text{elec}}$. Hence

$$Z_{\text{tot}} = \sum e^{-\beta(E_{\text{trans}} + E_{\text{rot}} + E_{\text{vib}} + E_{\text{elec}})}.$$

Extending over every possible quantum state of the assembly, the indicated summation must then extend over every possible *combination* of the assembly's translational, rotational, vibrational, and electronic quantum states. In these circumstances an argument basically identical with that used in the derivation of equation (10) here produces

$$Z_{\text{tot}} = \sum e^{-\beta E_{\text{trans}}} \times \sum e^{-\beta E_{\text{rot}}} \times \sum e^{-\beta E_{\text{vib}}} \times \sum e^{-\beta E_{\text{elec}}}.$$

But precisely these summations are what define the assembly partition functions for the *individual* degrees of freedom. We thus conclude from the last equation that

$$Z_{\text{tot}} = Z_{\text{trans}} \cdot Z_{\text{rot}} \cdot Z_{\text{vib}} \cdot Z_{\text{elec}}. \tag{16}$$

Calling now on equations (14) and (15), appropriate substitutions will at once yield

$$Z_{\text{tot}} = \left[\frac{1}{N!}(z_{\text{trans}})^N\right](z_{\text{rot}})^N(z_{\text{vib}})^N(z_{\text{elec}})^N. \tag{17}$$

For an assembly of *any* species of weakly coupled units, with *any* set of weakly coupled degrees of freedom, the last equation will permit us to express the assembly partition function Z in terms of the unit partition functions z. If the units are in free translational motion, we retain the bracketed term together with such other terms as refer to degrees of freedom that require consideration. Conversely, if the units are localized, we omit the first bracketed term, and we then retain only such other terms as refer to degrees of freedom requiring consideration. We have thus attained a formulation of the partition function that possesses great generality, but we must remain fully aware of severe restrictions on the domain of its applicability.

Bear in mind that assumptions of weak coupling underlie equation (17) and all the simple relations earlier derived in this section. When the various degrees of freedom of a given unit are not weakly but strongly coupled, proper evaluation of partition functions may demand procedures more elaborate than those indicated above. And still further complications arise when the various units of a given assembly are not weakly but strongly coupled—as indeed they are in many of the systems (for example, liquids and solutions) of greatest current interest to statistical mechanicians. An adequate account of such assemblies demands use of a statistical mechanics which—though its mathematical formalism remains much the same—must be founded on the rather different style of analysis first suggested by J. W. Gibbs. Happily, there remains ample scope for application of the statistical mechanics we have been developing on the premise of weak coupling.

THERMODYNAMIC PARAMETERS FROM PARTITION FUNCTIONS

Having in hand a general formulation of the partition function Z, we now show how all the essential thermodynamic parameters of a macroscopic assembly can be expressed in terms of Z.

Energy. Multiplying the number of units in each quantum state by the energy characteristic of that state, we can draw the total energy of an assembly of weakly coupled units from the very simple summation

$$E = \eta_0\epsilon_0 + \eta_1\epsilon_1 + \eta_2\epsilon_2 + \cdots \tag{e}$$

Equation (9) supplies appropriate expressions for the various population numbers, substitution of which yields

$$E = \epsilon_0 \frac{N}{z} e^{-\beta\epsilon_0} + \epsilon_1 \frac{N}{z} e^{-\beta\epsilon_1} + \epsilon_2 \frac{N}{z} e^{-\beta\epsilon_2} + \cdots,$$

where z symbolizes the usual partition function *per unit*. A slight further manipulation produces

$$E = \frac{N}{z}(\epsilon_0 e^{-\beta\epsilon_0} + \epsilon_1 e^{-\beta\epsilon_1} + \epsilon_2 e^{-\beta\epsilon_2} + \cdots) = N\frac{\sum \epsilon_q e^{-\beta\epsilon_q}}{\sum e^{-\beta\epsilon_q}}.$$

Treating all the ϵ_q terms as constants, we easily see how each of the additive terms in the numerator can be obtained as the negative derivative of the corresponding additive term in the denominator. That is

$$-\frac{d}{d\beta}e^{-\beta\epsilon_i} = \epsilon_i \cdot e^{-\beta\epsilon_i}.$$

Hence

$$E = -N\frac{\dfrac{d}{d\beta}\sum e^{-\beta\epsilon_q}}{\sum e^{-\beta\epsilon_q}} = -N\frac{1}{z}\frac{dz}{d\beta} = -N\frac{d\ln z}{d\beta}.$$

Moving the constant N inside the derivative, we move it still further within the logarithm, to find

$$E = -\frac{d(N\ln z)}{d\beta} = -\frac{d\ln z^N}{d\beta}.$$

Since for distinguishable units $z^N = Z$, it follows that

$$E = -\frac{d\ln Z}{d\beta}.$$

As so calculated, E will of course be based on whatever reference zero of energy was used in formulating Z—which is most often defined by writing $\epsilon_0 \equiv 0$.

Will the same equation for E apply also to assemblies of indistinguishable units—for example, to gas molecules in free translational motion? Let us see what happens when we substitute for Z the expression appropriate for such units. We then find

$$E = -\frac{d\ln\dfrac{1}{N!}z^N}{d\beta} = -\frac{d}{d\beta}\ln\frac{1}{N!} - \frac{d}{d\beta}\ln z^N.$$

But, with N a constant for the given assembly, the first term on the right reduces to zero. All that then remains is

$$E = -\frac{d\ln z^N}{d\beta} = -N\frac{d\ln z}{d\beta},$$

and from this we can easily retrace our steps to equation (e). And having thus recovered our original expression for E, we may be sure that the equation $E = -d \ln Z/d\beta$ applies alike to both localized units and to units in free translational motion. However the application to gas molecules *does* require that we add one small qualification to that equation.

In writing

$$-\frac{d}{d\beta} e^{-\beta\epsilon_i} = \epsilon_i \cdot e^{-\beta\epsilon_i}$$

we treated as a *constant* each of the energy terms here represented by ϵ_i. However, we will soon find that the energies associated with the translational quantum states are expressed by an equation of the form $\epsilon_i = f(1/V)^{2/3}$, where V symbolizes the volume of the assembly. Thus these energies are functions of a volume that will ordinarily vary in response to the change of temperature implied by differentiation with respect to β. Fortunately, any such variation of the energy terms can easily be avoided, simply by imposing a *constant-volume* restriction we indicate with a subscript V as we write:

$$E = -\left[\frac{d \ln Z}{d\beta}\right]_V. \tag{f}$$

Once established, the proposition that $\beta = 1/kT$ will entail that

$$d\beta = \frac{-dT}{kT^2} \quad \text{or} \quad \frac{dT}{d\beta} = -kT^2.$$

Calling on the chain rule, we rewrite equation (f) as

$$E = -\left[\frac{d \ln Z}{dT}\right]_V \left[\frac{dT}{d\beta}\right]_V.$$

Since neither T nor β is a function of V, $[dT/d\beta]_V = dT/d\beta$, so that

$$E = +kT^2 \left[\frac{d \ln Z}{dT}\right]_V. \tag{18}$$

Here then is our final expression for E as $f(Z)$.

Entropy. In Chapter 1 the statistical concept of entropy was expressed in the definition

$$S \equiv k \ln W.$$

For distinguishable units we know that

$$W = N!/\prod \eta_q!.$$

What will we then obtain for $\ln W$? Obviously

$$\ln W = \ln N! - \ln \prod \eta_q! = \ln N! - \sum \ln \eta_q!.$$

Application of Stirling's approximation will then yield

$$\ln W = N \ln N - N - \sum \eta_q \ln \eta_q + \sum \eta_q.$$

Of course the summation over all quantum states makes $\sum \eta_q = N$, so that

$$\ln W = N \ln N - \sum \eta_q \ln \eta_q.$$

Again drawing on equation (9), substitution now gives

$$\ln W = N \ln N - \sum \eta_q \ln \frac{N}{z} e^{-\beta \epsilon_q}$$

$$= N \ln N - \sum \eta_q \ln \frac{N}{z} - \sum \eta_q(-\beta \epsilon_q).$$

For all terms in the two summations the respective factors $\ln (N/z)$ and β are constants we can therefore move across the summation signs, to find

$$\ln W = N \ln N - \left(\ln \frac{N}{z} \right) \sum \eta_q + \beta \sum \eta_q \epsilon_q.$$

Again we set $\sum \eta_q = N$ and, by reference to equation (e), we easily recognize the second summation as an expression of the thermodynamic energy (E). Hence

$$\ln W = N \ln N - N \ln \frac{N}{z} + \beta E$$

$$= N \ln N - N \ln N + N \ln z + \beta E$$

$$= \ln z^N + \beta E.$$

Noting that for distinguishable units $z^N = Z$, we now draw on our definition of S to write

$$S \equiv k \ln W = k \ln Z + k\beta E.$$

So far derived only for distinguishable units, this equation can easily be shown to apply also to assemblies of identical gas molecules. To give this demonstration we cannot, of course, begin with $W = N!/\prod \eta_q!$, which applies only to distinguishable units. We must instead set out from equation (12), $W = \prod (\omega_y)^{n_y^*}/\prod \eta_y^*!$, which applies to identical units in free translational motion. Taking this as point of departure, **problem 15** invites use of equation (13) in a derivation much like the foregoing one. We so discover that

$$S \equiv k \ln W = -k \ln N! + k \ln z^N + k\beta E.$$

This differs from the result obtained just above only through the presence of the term $-k \ln N! = +k \ln (1/N!)$. Combining the logarithmic terms, we then find

$$S = k \ln \frac{1}{N!} z^N + k\beta E.$$

But the function inside the logarithm is *precisely* what we have found to represent the partition function Z for an assembly of units in free translational motion. For gas molecules as for localized units, we thus conclude again that

$$S = k \ln Z + k\beta E.$$

Once established, the proposition that $\beta = 1/kT$ will entail that

$$S = k \ln Z + \frac{E}{T}. \tag{19}$$

Substitution for E, from equation (18), permits us to rewrite the last equation wholly in terms of the partition function, as

$$S = k \ln Z + kT \left[\frac{d \ln Z}{dT} \right]_V = \frac{d}{dT} [kT \ln Z]_V.$$

Factorability and separability. Equations (18) and (19) provide expressions for the two fundamental thermodynamic functions, E and S. Expressions for all other thermodynamic functions can easily be obtained from these. For example, by definition the Helmholtz free energy $A \equiv E - TS$. Drawing on equation (19), we can then write

$$A = E - T \left(k \ln Z + \frac{E}{T} \right) = E - kT \ln Z - E.$$

And so we arrive at the remarkably simple result

$$A = -kT \ln Z. \tag{20}$$

All such expressions for the thermodynamic parameters prove alike in one notable respect: in every case the parameter is expressed as some function of the natural *logarithm* of Z. From this circumstance we can extract a highly significant message: our earlier demonstration that Z is *factorable* now immediately implies that terms in $\ln Z$ will be *separable*.

As an example, consider the last expression obtained for the entropy. When the units forming an assembly have multiple weakly coupled degrees of freedom, equation (16) will permit us to write:

$$S_{tot} = \frac{d}{dT} [kT \ln Z_{tot}]_V = \frac{d}{dT} [kT \ln Z_{trans} \cdot Z_{rot} \cdot Z_{vib} \cdot Z_{elec}]_V$$

$$= \frac{d}{dT} [kT \ln Z_{trans}]_V + \frac{d}{dT} [kT \ln Z_{rot}]_V$$

$$+ \frac{d}{dT} [kT \ln Z_{vib}]_V + \frac{d}{dT} [kT \ln Z_{elec}]_V$$

$$= S_{trans} + S_{rot} + S_{vib} + S_{elec}.$$

Here the last terms on the right respectively symbolize the entropy *contribution* arising from the translational degree of freedom, the corresponding *contribution* arising from the rotational degree of freedom, and all the other *contributions* that together make up the total value of the entropy. When the various degrees of freedom are only weakly coupled, so that the contribution of each is separately determinable, a simple act of addition then yields the value of the total entropy. But, beyond such convenience in calculation, we draw from this circumstance a unique power to *explain* the magnitudes of such thermodynamic parameters as the entropy. Thus we may find it possible to explain the unusually large entropy of some particular substance by pointing, for example, to a rotational contribution that, for assignable reasons, looms very large in this particular instance.

Evaluation of Partition Functions

3

Knowing how to express the thermodynamic parameters in terms of the partition function Z, we are still in no position to determine those parameters. For though we well know how to *formulate* Z, we have yet to learn how to *evaluate* Z. To such evaluations we now turn.

TRANSLATIONAL PARTITION FUNCTION

To the translational motion of a mass m, with total velocity u, classical physics assigns a kinetic energy (ϵ) given by the equation $\epsilon = \frac{1}{2}mu^2$. The total velocity can of course be resolved into velocity components $(u_x, u_y,$ and $u_z)$ respectively paralleling the axes of a Cartesian coordinate system. By the Pythagorean theorem

$$\epsilon = \frac{1}{2}mu^2 = \frac{m}{2}(u_x^2 + u_y^2 + u_z^2).$$

Since in classical physics all the velocity components, and u itself, are continuous variables, ϵ also is continuously variable. In quantum physics this is no longer true, and a simple particle-in-a-box treatment yields the relevant quantization condition. For a minute body confined to a rectangular box with dimensions L_x, L_y, and L_z, we find that the permissible energies are reduced to the discontinuous set of specific values given by the equation

$$\epsilon = \frac{h^2}{8m}\left[\frac{n_x^2}{L_x^2} + \frac{n_y^2}{L_y^2} + \frac{n_z^2}{L_z^2}\right],$$

where n_x, n_y, and n_z are translational quantum numbers that can assume only the *integral* values 1, 2, 3, ... and on up. Observe that, just as in the classical case, the total kinetic energy is resolvable into components severally associated with component motions paralleling the x-, y-, and z-axes.

Let us pausè a moment to develop two points of retrospective interest. For a particle in a cubic box with edge length L, $L_x^2 = L_y^2 = L_z^2 = L^2 = V^{2/3}$, where V symbolizes the volume of the container. The last equation then reduces to

$$\epsilon = \frac{h^2}{8mV^{2/3}} [n_x^2 + n_y^2 + n_z^2].$$

Observe first how easily degeneracy arises in the translational degrees of freedom: *different* quantum states, characterized by *different* values of n_x, n_y, and n_z, will all have the *same* energy whenever $n_x^2 + n_y^2 + n_z^2$ is the *same*. Observe second how the energy associated with any translational quantum state depends on the volume of the system. This dependence is what prompted us, on several earlier occasions, to impose a constant-volume restriction to ensure constancy of the indicated energies.

Now proceeding onward, we narrow our focus to center on an ideal monatomic gas. This is a system perfectly suited to our study, since any findings made for this assembly can readily be confirmed by comparison with empirical data for the noble gases—which are available real gases that *excellently* approximate ideal monatomic gases. Of course a monatomic gas has neither rotational nor vibrational degrees of freedom and, if we assume the absence of electronic excitations, we can exclude *all* internal degrees of freedom. Our choice of an ideal *monatomic* gas thus yields an attractively simple problem: the assignment of thermodynamic properties for an assembly of identical units that have *only* translational degrees of freedom. However, given the element of separability noted in the last section, our results will have a much broader significance. For they will also represent the translational *contribution* to the thermodynamic properties of ideal *polyatomic* gases, in which the units possess both translational *and* internal degrees of freedom. When, in a later section, we come to consider such polyatomic gases, we will then find further use for the results now to be worked out.

Ideal monatomic gas. For the energy (ϵ_x) associated with the x-component of motion of an ideal-gas molecule, our earlier analysis yields

$$\epsilon_x = \frac{n_x^2 h^2}{8mL_x^2}.$$

For the components of motion parallel to the y- and z-axes we can write similar expressions respectively associating ϵ_y and ϵ_z with n_y and n_z. Since we can regard these motions in three dimensions as three independent degrees of translational freedom, we can *separately* evaluate partition functions for each of them. The partition function (z_{trx}) for the x-component of motion will be

$$z_{trx} = \sum e^{-\beta n_x^2 h^2 / 8mL_x^2}$$

where the summation extends over all positive integral values of the quantum number n_x.

Again we remark that the accessible translational quantum states are both exceedingly numerous and *very* narrowly separated in energy. As indicated in **problem 17**, the summation over such a virtual continuum of states can be excellently approximated as an integral. We have then

$$z_{trx} = \int_0^\infty e^{-\beta n_x^2 h^2 / 8 m L_x^2} \, dn_x.$$

Defining $\alpha^2 \equiv \beta h^2 / 8 m L_x^2$, we recast the integral in the form $\int_0^\infty e^{-\alpha^2 u^2} \, du$. A straightforward mathematical analysis (see **problem 17**) yields for this well-known definite integral the result $\sqrt{\pi}/2\alpha$. Consequently

$$z_{trx} = \frac{\sqrt{\pi}}{2\sqrt{\beta h^2 / 8 m L_x^2}} = \frac{L_x}{2h} \left[\frac{8\pi m}{\beta} \right]^{1/2} = \frac{L_x}{h} \left[\frac{2\pi m}{\beta} \right]^{1/2}.$$

Evaluations of z_{try} and z_{trz} yield corresponding expressions in terms of L_y and L_z respectively. For the complete translational partition function (z_{trans}) we can then write

$$z_{trans} = z_{trx} \cdot z_{try} \cdot z_{trz} = \left[\frac{2\pi m}{\beta} \right]^{3/2} \cdot \frac{L_x L_y L_z}{h^3}.$$

Or, since the product of the linear dimensions of the box must equal the volume (V) thereof,

$$z_{trans} = \left[\frac{2\pi m}{h^2 \beta} \right]^{3/2} V.$$

Now at last we can supply the long-promised demonstration that $\beta = 1/kT$, where T symbolizes a temperature easily measurable on the familiar ideal-gas scale. On the strength of equation (14), we substitute in equation (f) on p. 60, to find

$$E = -\left[\frac{d \ln Z_{trans}}{d\beta} \right]_V = -\frac{d}{d\beta} \left[\ln \frac{1}{N!} (z_{trans})^N \right]_V.$$

Substitution for z_{trans}, from the last equation of the preceding paragraph, then yields

$$E = -\frac{d}{d\beta} \left[\ln \frac{1}{N!} \left(\frac{2\pi m}{h^2 \beta} \right)^{3N/2} \cdot V^N \right]_V$$

$$= -\frac{d}{d\beta} \left[\ln \frac{1}{N!} \left(\frac{2\pi m}{h^2} \right)^{3N/2} \cdot V^N \right]_V - \frac{d}{d\beta} \ln \left(\frac{1}{\beta} \right)^{3N/2}_V.$$

The square bracket contains only terms that (under the constant-volume restriction) are all *constants*. Hence the first derivative reduces to zero, and all that then remains is

$$E = +\frac{3N}{2} \frac{d}{d\beta} \ln \beta = \frac{3N}{2\beta}.$$

For one mole of units, the definition of the heat capacity (C_V) yields

$$C_V \equiv \left[\frac{d\bar{E}}{dT}\right]_V = \frac{d}{dT}\left[\frac{3N}{2\beta}\right]_V = \frac{3N}{2}\frac{d}{dT}\left(\frac{1}{\beta}\right).$$

But from experiment we know that at constant volume (and pressures low enough to ensure a close approach to ideality) the molar heat capacity of a monatomic noble gas is accurately expressible as

$$C_V = 12.47 \text{ joule/mole-}°K = 1.247 \times 10^8 \text{ erg/mole-}°K.$$

Consequently

$$\frac{3N}{2}\frac{d}{dT}\left(\frac{1}{\beta}\right) = 1.247 \times 10^8 \text{ erg(mole-}°K)^{-1}$$

$$\frac{d}{dT}\left(\frac{1}{\beta}\right) = \frac{2}{3}\frac{1.247 \times 10^8 \text{ erg(mole-}°K)^{-1}}{6.023 \times 10^{23} \text{ (mole)}^{-1}} = 1.38 \times 10^{-16} \text{ erg/}°K.$$

Now when first we remarked that β is an inverse function of temperature only, we defined an absolute statistical scale of temperature (θ) by writing $\theta \equiv 1/k\beta$. Here k is an as yet unspecified universal *constant* with the dimensions of energy/°K. Hence $1/\beta = k\theta$, and substitution in the last equation of the preceding paragraph will give

$$k\frac{d\theta}{dT} = 1.38 \times 10^{-16} \text{ erg/}°K.$$

Let us then assign to k the value 1.38×10^{-16} erg/°K, for by this assignment we will fully ensure that

$$d\theta = dT.$$

That is, any temperature *interval* on the absolute θ-scale will be equal to the corresponding temperature *interval* on the ideal-gas T-scale. But earlier (on p. 33) we found good reason to suppose that 0° on the θ-scale is also the temperature represented by 0° on the T-scale. Thus we can at once integrate the last equation, to find

$$\int_0^\theta d\theta = \int_0^T dT,$$

$$\theta = T.$$

And so we are led to conclude that the absolute statistical θ-scale of temperature may be made *everywhere* perfectly coincident with the familiar and fully operational ideal-gas T-scale of temperature.

Returning then to our consideration of the partition function of an ideal monatomic gas, where before we had obtained only $z_{trans} =$

$[2\pi m/h^2\beta]^{3/2}V$, now we can write

$$z_{\text{trans}} = \left[\frac{2\pi mkT}{h^2}\right]^{3/2} V. \tag{21}$$

By virtue of equation (14), the corresponding expression for an N-unit assembly will be

$$Z_{\text{trans}} = \frac{1}{N!}\left\{\left[\frac{2\pi mkT}{h^2}\right]^{3/2} V\right\}^N. \tag{a}$$

And where we had before obtained only $E = 3N/2\beta$, now we can write

$$E = \tfrac{3}{2}NkT. \quad = \tfrac{3}{2}RT - \text{for mole}$$

Finally, for one mole of ideal monatomic gas, we can now write

$$C_V = \tfrac{3}{2}Nk. \quad = \tfrac{3}{2}R. \quad - \text{for mole}.$$

The last equation implies that C_V is wholly independent of temperature. At the lowest temperature to which experimental measurements can be extended, monatomic gases do indeed retain the indicated heat capacity. However, we must not push this conclusion *too* far—if only because the equation $dS = C_V \, dT/T$ would blow up in the region where $T \to 0$. In this region the sharp contraction in the number of accessible translational quantum states vitiates the "dilute gas" approximation and, as noted on p. 56, a rather more sophisticated analysis is then required. This does indeed demonstrate that in the limit $T \to 0$ so also does $C_V \to 0$.

We turn now to the thermodynamic function most easily expressed in terms of the partition function. Substituting from equation (a) in equation (20), we find for the Helmholtz free energy of an ideal monatomic gas

$$A = -kT \ln Z = -kT \ln \frac{1}{N!} z^N = -kT[-\ln N! + N \ln z].$$

Application of Stirling's approximation then yields

$$A = -kT[-(N \ln N - N) + N \ln z] = -NkT(1 - \ln N + \ln z)$$

whence it follows that

$$A = -NkT\left(1 + \ln \frac{z}{N}\right). \tag{b}$$

This apparently unilluminating relation at once gives rise to an unforeseen conclusion. The definition $A \equiv E - TS$ implies

$$dA = dE - T \, dS - S \, dT$$

This may be joined with a familiar thermodynamic relation, $dE = T \, dS - P \, dV$, to yield

$$dA = -P \, dV - S \, dT.$$

This is a *general* relation, subject only to the restriction that $P\,dV$ work be the only species of work possible. If in addition we stipulate constant temperature, then $dT = 0$, and the thermodynamic relation is reduced to:

$$P = -\left[\frac{dA}{dV}\right]_T.$$

Substituting from equation (b), we now express P in terms of the partition function:

$$P = -\frac{d}{dV}\left[-NkT\left(1 + \ln\frac{z}{N}\right)\right]_T.$$

Since for this differentiation T is stipulated to be a constant, and since for any given assembly N is a constant, the $-NkT$ term may be removed from within the differential operator, leaving the figure 1 as a constant with derivative = 0. Substituting for z from equation (21), we have then:

$$P = +NkT\frac{d}{dV}\left[\ln\frac{z}{N}\right]_T = NkT\frac{d}{dV}\left\{\ln\left[\frac{2\pi mkT}{h^2}\right]^{3/2}\frac{V}{N}\right\}_T$$

$$= NkT\frac{d}{dV}\left\{\ln V + \ln\frac{1}{N}\left[\frac{2\pi mkT}{h^2}\right]^{3/2}\right\}_T$$

$$= NkT\frac{d}{dV}[\ln V]_T + NkT\frac{d}{dV}\left\{\ln\frac{1}{N}\left[\frac{2\pi mkT}{h^2}\right]^{3/2}\right\}_T.$$

For a given assembly of units at constant temperature, *all* the terms in the last derivative are constants. Hence that derivative reduces to zero, leaving only:

$$P = NkT\frac{d}{dV}[\ln V]_T = NkT\frac{1}{V}.$$

For one mole of units, N becomes Avogadro's number N, in which case $Nk = \mathbf{N}k = 6.023 \times 10^{23}$ (mole)$^{-1}$ $\times 1.38 \times 10^{-16}$ erg/°K = 8.314×10^7 erg/mole-°K = 8.314 joule/mole-°K = 0.08205 lit-atm/mole-°K. But this is precisely the numerical constant we ordinarily symbolize by the letter R. And so we have derived for one mole of ideal monatomic gas the familiar equation

$$PV = RT.$$

However familiar the equation, its appearance at the end of the above derivation is an electrifying development. The average reader is likely to have encountered this equation on three earlier occasions. First, as a generalization of the results of experiments on gases—but here we have said nothing of experiments. Second, as a conclusion drawn by deduction from a kinetic theory of gases that conceives "pressure" as resulting from "molecular impact"—but we have here said nothing about molecular impact. Third, as an explicit definition required for the application of thermodynamics to a particular kind of system—but here we have needed no such definition. For, in the present derivation, we have obtained the ideal-gas law purely as a consequence of the quantization condition for the translational state of structureless particles. Our derivation thus symbolizes a typical capacity of statistical mechanics: the capacity to *produce* relations that appear in thermodynamics only as empirical generalizations and/or definitions. It may be objected that in this derivation we drew on thermodynamics for the equation $dA = -P\,dV - S\,dT$. However this was more a matter of convenience than of necessity, for the ideal-gas law is also derivable by a more elaborate (but purely statistical) analysis that makes no call whatever on the relations of thermodynamics. Another possible objection is that our derivation may be tainted by circularity. For, in earlier establishing that $\beta = 1/kT$, we certainly drew somewhat on prior knowledge of the behavior of ideal *monatomic* gases. However, we can now easily show that the very same law applies also to *polyatomic* gases, about which nothing has so far been assumed.

For a monatomic gas, possessing *only* translational degrees of freedom, equation (b) yields the *total* value of the Helmholtz free energy. With a polyatomic gas, other contributions to the free energy must of course be considered. But, from the last chapter, we already know how these contributions will enter into the total value of the free energy:

$$A_{\text{tot}} = A_{\text{trans}} + A_{\text{rot}} + A_{\text{vib}} + \cdots$$

Two remarks are now in order. First, for a polyatomic gas the translational contribution will be identical with the total value of A computed for an ideal monatomic gas. Why? Because the quantization of the translational states remains exactly the same whether or not the units have available to them other *independent* degrees of freedom. Second, anticipating results to be obtained presently, we note that the partition functions for rotation, vibration, etc., contain no term in V or in any parameter that is a function of V. Hence $A_{\text{rot}}(= -kT \ln Z_{\text{rot}})$ and $A_{\text{vib}}(= -kT \ln Z_{\text{vib}})$ are functions invariant with V. Expressing the pressure of a polyatomic gas in terms of A_{tot}, we have then:

$$P = -\left[\frac{dA_{\text{tot}}}{dV}\right]_T = -\frac{d}{dV}\left[-NkT\left(1 + \ln\frac{z}{N}\right) + A_{\text{rot}} + A_{\text{vib}} + \cdots\right]_T$$

$$= \frac{d}{dV}\left[NkT\left(1 + \ln\frac{z}{N}\right)\right]_T - \left[\frac{dA_{\text{rot}}}{dV}\right]_T - \left[\frac{dA_{\text{vib}}}{dV}\right]_T - \cdots$$

The first term remains identical with that found for a monatomic gas, and will yield the identical result $[= NkT(1/V)]$. And all succeeding terms represent derivatives, with respect to V, of functions that contain no term in V or in any parameter that is a function of V. So far as this differentiation is concerned, all these functions are constants, and all the corresponding derivatives reduce to zero. Thus we obtain for a polyatomic ideal gas exactly the same equation earlier derived only for an ideal monatomic gas:

$$P = NkT \frac{1}{V}, \quad \text{or, for one mole of gas,} \quad PV = RT.$$

Turning to one last major thermodynamic function, we observe that, since $A \equiv E - TS$, the entropy $S = (E - A)/T$. But we have already derived expressions for the E and A of ideal monatomic gases. Substituting these, we find:

$$S = \frac{E - A}{T} = \frac{\frac{3}{2}NkT + NkT[1 + \ln (z/N)]}{T} = Nk \left(\frac{5}{2} + \ln \frac{z}{N} \right).$$

Substitution for z, from equation (21), then yields:

$$S = Nk \left\{ \frac{5}{2} + \ln \left[\frac{2\pi mkT}{h^2} \right]^{3/2} \frac{V}{N} \right\}.$$

For one mole of material we can write $Nk = \mathbf{N}k = R$, and we also multiply the numerator and denominator of the logarithmic term by $(\mathbf{N})^{3/2}$. We can then replace the grouping $(\mathbf{N}m)^{3/2}$ by $M^{3/2}$—where M represents the molecular weight of the gas concerned—and thus acquire the expression:

$$\bar{S} = R \left\{ \frac{5}{2} + \ln \left[\frac{2\pi MkT}{h^2 \mathbf{N}} \right]^{3/2} \frac{V}{\mathbf{N}} \right\}.$$

Some interesting points may be brought out by systematic rearrangement to:

$$\bar{S} = R \ln V + \frac{3}{2} R \ln T + \frac{3}{2} R \ln M + \left\{ R \left(\frac{5}{2} + \ln \left[\frac{2\pi k}{h^2 \mathbf{N}^{5/3}} \right]^{3/2} \right) \right\}. \quad \text{(c)}$$

The last term, in braces, is nothing but a constellation of *constants*, actual substitution for which results in the figure -11.074 cal/mole-°K.

Equation (c) is the celebrated Sackur-Tetrode equation. What do we learn from it? The first term represents no novelty. Taking one mole of ideal monatomic gas at some volume V_1, suppose we ask how its entropy changes when, *at constant temperature*, its volume changes to V_2. Using equation (c) to express $\bar{S}_1$ and $\bar{S}_2$, we write $\Delta \bar{S} = \bar{S}_2 - \bar{S}_1$. But then only the first of the four terms on the right of equation (c) will survive the indicated subtraction, yielding:

$$\Delta \bar{S} = R \ln V_2 - R \ln V_1 = R \ln \frac{V_2}{V_1}.$$

But this is a perfectly well-known equation in thermodynamics. The second term of the Sackur-Tetrode equation is also no novelty. Taking one mole of ideal monatomic gas at some temperature T_1, suppose we ask how its entropy changes when, *at constant volume*, its temperature changes to T_2. Using equation (c) to express $\bar{S}_1$ and $\bar{S}_2$, we write $\Delta\bar{S} = \bar{S}_2 - \bar{S}_1$. But then only the second of the four terms on the right of equation (c) will survive the indicated subtraction, so that:

$$\Delta\bar{S} = \frac{3}{2} R \ln T_2 - \frac{3}{2} R \ln T_1 = \frac{3}{2} R \ln \frac{T_2}{T_1}.$$

Having already established that $\frac{3}{2}R = C_V$ for an ideal monatomic gas, we have then:

$$\Delta\bar{S} = C_V \ln \frac{T_2}{T_1}.$$

This too is a perfectly well-known equation in thermodynamics. Even the third term in the Sackur-Tetrode equation is incompletely a novelty: this term had already been suggested by Sackur, on strictly empirical grounds, before the Sackur-Tetrode equation was derived. Only the fourth term in that equation is a complete novelty—showing us how to calculate, from a constellation of constants, a number that could previously have been estimated only on empirical grounds.

What the Sackur-Tetrode equation does thus comes to this. It offers, as the products of statistical analysis, two terms previously well known from thermodynamics, and it embodies two additional terms that classical thermodynamics is quite powerless to produce. We now have *two* radically different ways of evaluating the entropies of ideal monatomic gases. From measurements of heat capacity, at various temperatures, we can make an *empirical* determination of entropies. And we can also make a *theoretical* calculation of those entropies—by using the Sackur-Tetrode equation which, for the present purpose, may best be written in what **problem 19** shows to be an equivalent form when pressure is expressed in atmospheres:

$$\bar{S}_{\text{trans}} = R \ln M^{3/2} T^{5/2}/P - 2.316 \text{ cal/mole-}°\text{K}. \tag{22}$$

The *excellent* agreement of empirically and theoretically determined entropies, in cal/mole-°K for gases at 298°K and 1-atm pressure, is displayed in the accompanying tabulation, after K. K. Kelley. Observe that, so great is the conviction aroused by the statistical derivation, the empirical values are regarded as *less* reliable than those calculated theoretically.

	Calorimetric	*Theoretical*
Ne	35.01 ± 0.1	34.95 ± 0.01
Ar	36.95 ± 0.2	36.99 ± 0.01
Kr	39.17 ± 0.1	39.20 ± 0.01
Xe	$40.7 \ \pm 0.3$	40.54 ± 0.01

The multitude of translational states. We come now to the long-promised demonstration that, in ordinary circumstances, the accessible translational quantum states are so exceedingly numerous that the vital "dilute gas" condition will be fully satisfied. Opening an illuminating perspective on the results obtained in the last section, this demonstration serves also to lay the foundation for a derivation (in the next section) of the Maxwell-Boltzmann molecular-speed distribution.

For a molecule of mass m in a volume V, we found (on p. 65) the energy-quantization condition

$$\epsilon = \frac{h^2}{8mV^{2/3}} [n_x^2 + n_y^2 + n_z^2],$$

where n_x, n_y, and n_z are integral quantum numbers. Let us now define $n^2 \equiv n_x^2 + n_y^2 + n_z^2$, where n is an index number that need *not* be integral. We thus attain a very compact expression for the translational energy, as

$$\epsilon = \frac{n^2 h^2}{8mV^{2/3}}.$$

Recall now that (on p. 68) we obtained, for the translational energy possessed by N gas molecules, the result

$$E = \tfrac{3}{2}NkT.$$

The quotient E/N then yields for the *average* energy per molecule

$$\epsilon = \tfrac{3}{2}kT.$$

Equating the last two expressions for ϵ, we at once obtain a relation from which we can calculate the *average* value of the index number n:

$$\frac{n^2 h^2}{8mV^{2/3}} = \frac{3}{2} kT$$

$$n^2 = \frac{12mkT}{h^2} V^{2/3}.$$

With a view to forthcoming developments, we slightly recast the numerical coefficient, and then solve for both n and n^3:

$$n^2 = \frac{6}{\pi} \frac{2\pi mkT}{h^2} V^{2/3},$$

$$n = \left[\frac{6}{\pi} \frac{2\pi mkT}{h^2} \right]^{1/2} \cdot V^{1/3} \quad \text{and} \quad n^3 = \left[\frac{6}{\pi} \right]^{3/2} \left\{ \left[\frac{2\pi mkT}{h^2} \right]^{3/2} V \right\}.$$

Apart from the numerical coefficient, the last function is identical with that obtained, in equation (21), for the partition function z_{trans}. To bring out the

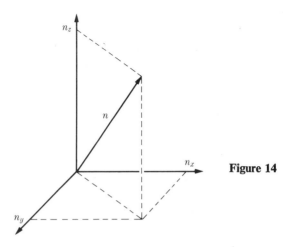

Figure 14

significance of this near-identity, we now relate the term n^3 to the total number of translational quantum states accessible to the gas molecule.

Imagine a coordinate space in which we plot the quantum numbers n_x, n_y, and n_z along three mutually perpendicular Cartesian axes. Each possible quantum state—characterized by some set of positive values of n_x, n_y, and n_z—is then represented by a single point falling in the positive octant of the coordinate space, which is thus "filled" by points with a uniform unit spacing that reflects the requirement that n_x, n_y, and n_z shall be *integral*. By the Pythagorean theorem, in this coordinate-space the index number $n(\equiv \sqrt{n_x^2 + n_y^2 + n_z^2})$ corresponds to the radial length indicated in Fig. 14. That is, the index-number n symbolizes the radius of a sphere the positive octant of which contains all points that represent quantum states with energies ranging up to a maximum set by the selected value of n. By counting the number of such points, let us now determine the number of quantum states having energies ranging up to the maximum established by the value of n calculable from the expression on the preceding page. The apparently difficult count of points proves unexpectedly simple as soon as one finds (see **problem 18**) that, under ordinary circumstances, this average value of n is a *huge* number.

When "points" fall at unit spacing in a uniform three-dimensional lattice, their number is very simply related to the lattice volume as indicated in Fig. 15—where, to avoid confusion, we have shown only those points visible on the outside surface of the lattice. Symbolizing by l the number of units of edge-length for a cubic portion of the lattice, we see that the volume will be l^3 while the number of points is $(l + 1)^3$. Now when n is a large number we will be concerned with a very large chunk of the lattice, and in the

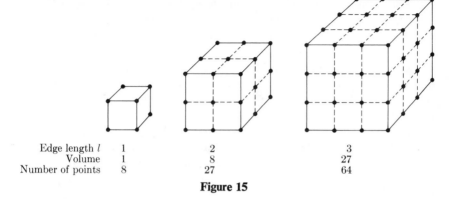

Edge length l	1	2	3
Volume	1	8	27
Number of points	8	27	64

Figure 15

limit of very large l clearly the number of points present can be *excellently* approximated by the volume of the lattice. To determine the number of points comprised within the positive octant of a sphere of radius n, we have then only to determine the volume of that spherical octant. This we can easily do.

Symbolizing by ζ the number of quantum states represented by those points, we will write

$$\zeta = \frac{1}{8} \times \frac{4}{3} \pi n^3 = \frac{\pi}{6} n^3.$$

Up to a maximum that corresponds to the *average* energy of the gas molecules, the value of n^3 is that given by the expression on p. 73. Substitution of this expression yields

$$\zeta = \frac{\pi}{6} \left[\frac{6}{\pi}\right]^{3/2} \left[\frac{2\pi m k T}{h^2}\right]^{3/2} V,$$

$$\zeta = 1.382 \left[\frac{2\pi m k T}{h^2}\right]^{3/2} V,$$

where we have replaced $\sqrt{6/\pi}$ by the numerically equivalent 1.382. For the representative example of a mole of neon (atomic weight, 20) at 1-atm pressure and 298°K, actual substitution of the appropriate numerical values yields for ζ a perfectly gigantic figure, of the order of 10^{30}.

Observe that 10^{30} is no more than a *minimum* estimate, since we have counted only those quantum states with energies ranging up to the average value of $\frac{3}{2}kT$. To exactly the extent that some molecules fall short of that average energy, other molecules must possess energies in excess of the average—and so be in occupation of quantum states we have *not* counted. But, even on a minimum estimate of ζ, with no fewer than 10^{30} quantum states uniformly distributed through the n_x-n_y-n_z coordinate space, clearly the energy *difference* between neighboring states will be utterly miniscule. Hence the assembly fully meets the two conditions stipulated when, in evaluating the translational partition function, we replaced the summation by an integral: (1) the accessible states *are* exceedingly numerous, and (2) the energy spacings *are* exceedingly minute. At a more fundamental level, we can now also demonstrate the "dilute gas" condition invoked when we first ventured to extend Boltzmann statistics to identical gas molecules in free translational motion. For, comparing the 10^{30} accessible quantum states with the 6×10^{23} molecules present in a mole of neon, we find that there are well over 10^6 times as many quantum states as there are molecules to be distributed over them. The molecules are then so thinly spread that the probability of multiple occupancy of those states is, as earlier assumed, effectively negligible.

We turn now to the near-identity of our expression for ζ and the expression earlier derived for the partition function z_{trans}. Why should the partition function be so nearly identical with the total number of accessible quantum states? The answer is not far to seek! When we conduct the sum-over-states that yields the partition function, we enter for each state a term $e^{-\epsilon_i/kT}$. As long as $\epsilon_i \ll kT$, the term has a value $\rightarrow e^{-0} = 1$. Later in the summation, when $\epsilon_i \rightarrow \frac{3}{2}kT$, the term has a value $\rightarrow e^{-1.5} \simeq 1/4.5 = 0.22$. Up to the average energy $\frac{3}{2}kT$, the partition-function summation will thus consist of ζ terms with values falling between a maximum of 1 and a minimum of 0.22. Allowing then for the accessibility of states with energies somewhat above the average, we see that the partition-function summation will also include many additional terms with values falling between a maximum of 0.22 and a minimum $\rightarrow 0$. Observing that the presence of these later terms acts to compensate for the fact that some of the first ζ terms fall well short of unity, we may crudely approximate the partition function as the sum of ζ terms of the order of 1. That is, $z_{\text{trans}} \simeq \zeta$, which is indeed the conclusion implied by the near identity of the derived expressions for ζ and for z_{trans}.

The success of this order-of-magnitude interpretation of z_{trans} in terms of ζ suggests the feasibility of a similar intepretation of the Sackur-Tetrode equation for the entropy of a monatomic gas. In a *very* crude estimate of the number of ways in which a mole of gas molecules can be distributed over ζ quantum states, we entirely ignore the difference of the energies associated with those states. Grouping all ζ states into a *single* energy level with degeneracy ζ, we consider the number of microstates arising from the distribution of N identical molecules among those states. The formulation of this problem on p. 51 yields

$$W = \frac{\zeta^N}{N!}.$$

For one mole of gas molecules, therefore,

$$\bar{S} \equiv k \ln W = k \ln \zeta^N - k \ln N!$$

$$= Nk \ln \zeta - Nk \ln N + Nk$$

$$= R \ln \frac{\zeta}{N} + R.$$

Now the last expression for ζ given on p. 75 will permit us to write for one mole of ideal gas, described by the equation of state $P\bar{V} = RT$,[†]

$$\frac{\zeta}{N} = 1.382 \left[\frac{2\pi mkT}{h^2}\right]^{3/2} \cdot \frac{\bar{V}}{N} = 1.382 \left[\frac{2\pi MkT}{Nh^2}\right]^{3/2} \cdot \frac{RT}{PN}.$$

Here we have substituted for the particle weight (m) the quotient of the gram-molecular weight (M) divided by Avogadro's number (N). Substitution for ζ/N in the last equation of the preceding paragraph then gives

$$\bar{S} = R \ln \left\{\frac{1.38R}{N}\left[\frac{2\pi k}{Nh^2}\right]^{3/2}\frac{M^{3/2}T^{5/2}}{P}\right\} + R$$

$$= R \ln \frac{M^{3/2}T^{5/2}}{P} + R\left(1 + \ln \frac{1.38R}{N}\left[\frac{2\pi k}{Nh^2}\right]^{3/2}\right).$$

The first term on the right is identical with that in equation (22). And on making numerical substitutions for the indicated cluster of constants, we find for the second term a value (-4.65) of very much the same order of magnitude as the corresponding term in equation (22). For the entropy $\bar{S}$, as formerly for the partition function z_{trans}, we can thus base on ζ a thoroughly satisfactory order-of-magnitude interpretation. This interpretation is conspicuously successful in rationalizing the dependence of the entropy on the molecular weight (M) of the gas concerned. Recall that, as the energy-quantization condition for the translational degree of freedom, we found a function of the form $\epsilon = f(1/m)$. The larger the value of the particle weight (m), the narrower will be the energy spacing between the translational quantum states—and the greater will be the number of such states accessible at a given temperature. Thus molecules with a large molecular weight will be "spread" over a greater number of states than are accessible to molecules of small molecular weight, with consequent disparity of the molar entropies in the two cases.

The Maxwell-Boltzmann speed distribution. In a gas at a given temperature, all the molecules do not move with the same speed. Some move more slowly,

---

† If P is to be expressed in atmospheres then, as noted in **problem 18**, the appropriate value for *this* R will be 82.05 ml-atm/mole-°K.

others rather more rapidly. For the purposes of chemical kinetics (and for many other purposes as well) we wish to know what fraction of all the molecules present will have speeds falling within any particular range of values.

As usual when dealing with the translational degrees of freedom, we collect a great many quantum states together in each energy level, the population (n^*) of which is given by equation (13) as

$$\eta^* = \frac{N}{z}\, \omega e^{-mu^2/2kT}.$$

Apart from replacing $\beta\epsilon$ by ϵ/kT, we have further substituted for ϵ the equivalent expression ($\frac{1}{2}mu^2$) for the kinetic energy of a mass m that moves with the speed u. Now the large number (ω) of quantum states grouped together in the single energy level is but a minute portion ($d\zeta$) of the immense total number (ζ) of quantum states. In analogous fashion, the large population (η^*) of that energy level is but a minute portion (dN) of the immense total number (N) of units present. Hence we can properly reformulate the last equation as

$$\frac{dN}{N} = \frac{d\zeta}{z}\, e^{-mu^2/2kT},$$

where dN/N symbolizes the fraction of the molecules with a speed falling in a short range around the given value of u. Our only remaining problem is to find, in terms of u, a suitable expression for the degeneracy now symbolized by $d\zeta$.

In the n_x-n_y-n_z coordinate space, each translational quantum state is represented by a point. Up to any particular maximum energy, expressible in terms of an index number n, the total number of quantum states (ζ) was found on p. 75 to be

$$\zeta = \frac{1}{8}\left(\frac{4}{3}\pi n^3\right).$$

The number of quantum states ($d\zeta$) falling within any short span (dn) of the index number is then easily found by differentiation

$$d\zeta = \frac{1}{8}\left(\frac{4}{3}\,\pi 3n^2\right) dn$$

$$= \frac{1}{8}\,(4\pi n^2)\, dn = \frac{\pi}{2}\, n^2\, dn.$$

This result is easily understandable. When the index number increases from n to $n + dn$, the consequent increase in the radius of the sphere will somewhat increase the number of included quantum states or points. The increase ($d\zeta$) in the number of states is evidently equal to the number of points that fall in the annular region between the initial sphere with radius n and the final sphere with radius $n + dn$. This number of points is measured by the volume of the added region with thickness dn and area $\frac{1}{8}(4\pi n^2)$, that is, the surface of the positive octant of the original sphere. And this is precisely the result expressed in the last equation.

Having progressed from ω to $d\zeta$, and from $d\zeta$ to $(\pi/2)n^2 \, dn$, we now easily see how one further step will yield the sought-for expression of the degeneracy in terms of the molecular speed (u). For on p. 73 we have already obtained a relation that may be used to link the index number n to the translational kinetic energy now to be expressed as $\frac{1}{2}mu^2$. That is

$$\frac{n^2 h^2}{8mV^{2/3}} = \epsilon = \frac{1}{2} mu^2,$$

$$n^2 = \frac{4m^2 u^2 V^{2/3}}{h^2}.$$

Therefore

$$n = \frac{2mu V^{1/3}}{h} \qquad \text{and} \qquad dn = \frac{2mV^{1/3}}{h} \, du.$$

And now we *can* express the number of quantum states ($d\zeta$) associated with any small range of speeds (du) around any given speed u:

$$d\zeta = \frac{\pi}{2} n^2 \, dn = \frac{\pi}{2} \frac{8m^3 V}{h^3} u^2 \, du = 4\pi u^2 \cdot m^3 \cdot \frac{V}{h^3} \, du.$$

Return now to our equation for dN/N. Beyond substituting the above expression for $d\zeta$, we substitute also the expression for z that is supplied by equation (21). We thus arrive at

$$\frac{dN}{N} = \frac{4\pi u^2 \cdot m^3 \cdot \dfrac{V}{h^3} \, du}{\left[\dfrac{2\pi mkT}{h^2}\right]^{3/2} \cdot V} e^{-mu^2/2kT}.$$

Cancellation of V/h^3 from numerator and denominator, and slight rearrangement, then yields as the final result

$$\frac{1}{N} \frac{dN}{du} = 4\pi u^2 \left[\frac{m}{2\pi kT}\right]^{3/2} \cdot e^{-mu^2/2kT}.$$

This is the famous Maxwell-Boltzmann speed distribution law. What is symbolized by the function on the left? To make sense of the derivative, consider that if an object travels some short distance (ds) in some brief interval of time (dt), the quotient ds/dt will represent the distance traveled per *unit* time. In the same fashion, if some small number of molecules (dN) falls within some short range of speeds (du), the quotient dN/du will represent the number of molecules falling within a *unit* range of speed. Hence the function $(1/N)(dN/du)$ symbolizes the *fraction* of the molecules that fall within a range of 1 cm/sec around any particular speed u. And for any given speed u, we can calculate the corresponding fraction simply by substituting values for all terms on the right side of the last equation.

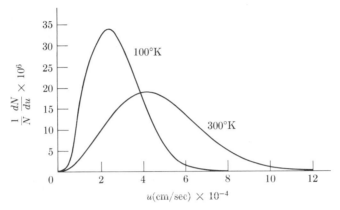

Figure 16

The variation of $(1/N)(dN/du)$ with u and T is displayed in Fig. 16 for nitrogen molecules at 100° and 300°K. At all comparatively low values of u, the function $e^{-mu^2/2kT} \to e^{-0} = 1$, and it is then the term $4\pi u^2$ that produces the steep rise as u increases from zero. This (parabolic) rise continues to higher values of u the higher the prevailing temperature. For where at low temperature a particular u may make $e^{-mu^2/2kT}$ significantly less than 1, at high temperature that same u may still yield $e^{-mu^2/2kT} \simeq e^{-0} = 1$. However, sooner at low temperature, and later at high temperature, the rise of $(1/N)(dN/du)$ with increasing u must come to an end. Why? Sooner or later the values of u become large enough that the parabolic rise consequent to the $4\pi u^2$ term will be outweighed by the exponential decline consequent to the $e^{-mu^2/2kT}$ term. As the exponential thus assumes control, the function $(1/N)(dN/du)$ passes through a maximum and then falls off toward zero. And this falling off will be the more abrupt the lower the temperature—since, the smaller the value of T, the smaller will be the value of u at which $e^{-mu^2/2kT} \to 0$. At high temperature, therefore, the molecules will be more thinly spread over a larger span of speeds, and the function $(1/N)(dN/du)$ will not attain the peak values achieved at lower temperatures. This effect is reflected in the more gradual rise of the high-temperature curve as u increases from zero. The apparent equality of the areas under both curves is also readily understandable: when summed over all values of u, the function $(1/N)(dN/du)$ must in both cases yield 1. Thus *every* qualitative feature of the Maxwell-Boltzmann speed distribution proves readily interpretable.

INTERNAL DEGREES OF FREEDOM

Having fully examined the translational degree of freedom, we turn now to the rotations, vibrations, and electronic excitations we class together as "internal" degrees of freedom.   Our earlier assumption of weak coupling here becomes decidedly vulnerable. Insofar as they alter force constants and internuclear distances, changes of electronic state certainly cannot fail to affect both the vibrational and rotational degrees of freedom.   And even when (as is most often the case) we can wholly ignore changes of electronic state, still we must recognize that the bare possibility of a molecular vibration is fundamentally incompatible with the usual *rigid*-rotor analysis we will employ.   Thus the postulated weak coupling of vibrational and rotational degrees of freedom can represent no more than an approximation.   However, this very convenient approximation is ordinarily quite good enough for our purposes, and on this basis we now proceed.

Rotation.   Conceiving a diatomic molecule as a rigid dumbbell-shaped object, we find in classical physics a very simple expression for its rotational kinetic energy:

$$\epsilon = \tfrac{1}{2}I\omega^2.$$

In this formula (which perfectly parallels the $\tfrac{1}{2}mu^2$ expression for translational kinetic energy) ω symbolizes the "angular velocity" defined as the number of radians swept out by the rotor in one second.   And I symbolizes the "moment of inertia" defined as $I \equiv \sum m_x r_x^2$, where r_x symbolizes the distance at which each atom with mass m_x stands from the axis of rotation. For a *diatomic* molecule a more useful equivalent expression is $I = \mu r^2$, where r symbolizes the internuclear distance and the so-called "reduced mass" is defined in terms of the two atomic masses (m_1 and m_2) by writing $\mu \equiv m_1 m_2/(m_1 + m_2)$.   When a given molecule rotates around a given axis, the value of I may be treated as effectively a *constant*.   But, since ω is continuously variable, a continuum of rotational energies is possible for the classical rigid rotor.   However, in quantum physics this is no longer true for a rotor of micro-cosmic dimensions.   An additional restriction then operates to reduce the permissible rotational energies to the noncontinuous set of discrete values given by the equation:

$$\epsilon = \frac{h^2}{8\pi^2 I} (J)(J + 1),$$

where J is the rotational quantum number which can assume any of the *integral* values 0, 1, 2, ...

Recalling our treatment of translational motions, we conceive molecular rotations as "resolved" into essentially independent component motions.

Just as every possible translational motion can be built up from component motions paralleling x-, y-, and z-axes, every possible rotational motion of a diatomic molecule can be built from component rotations around two mutually perpendicular axes intersecting at the molecular center of gravity. The above equation for the rotational energy thus expresses the total energy associated with the *two* rotational degrees of freedom of a *diatomic* molecule. To construct all possible rotations of a nonlinear *polyatomic* molecule, on the other hand, we must use component rotations around three "principal axes," and *three* degrees of freedom are thus associated with rotation of such a molecule. In that case the moments of inertia about the various axes may well be different, but this slight complication does not arise with the diatomic molecules that are our sole present concern.

Having found, on p. 46, that $(2J + 1)$ represents the degeneracy of the energy level for which the last equation expresses ϵ as $f(J)$, we formulate the rotational partition function as

$$z_{\text{rot}} = \sum \omega_J e^{-\epsilon_J/kT} = \sum (2J + 1)e^{-J(J+1)h^2/8\pi^2 IkT}. \tag{d}$$

For the ground state, with $J = 0$, the first term of the summation reduces to 1; and for the subsequent terms, with $J = 1, 2, \ldots$, we have:

$$z_{\text{rot}} = 1 + 3e^{-2h^2/8\pi^2 IkT} + 5e^{-6h^2/8\pi^2 IkT} + \cdots$$

This series, alas, does not sum to a closed analytical function. However, for molecules involving reasonably massive atoms (i.e., molecules with relatively large moment of inertia), at temperatures reasonably removed from $0°K$, the rotational energy levels are narrowly enough separated to be approximated as a continuum. The above summation can then be replaced by the corresponding integral. But, before undertaking this operation, let us make our symbology more compact by defining a new parameter θ_r:

$$\theta_r \equiv \frac{h^2}{8\pi^2 Ik}.$$

Having the dimensions of temperature, the parameter θ_r expresses for each substance the temperature at which the rotational levels above the ground state first begin to be appreciably populated. Observe that θ_r contains the *only* term (I) that reflects the identity of the substance for which a rotational partition function is to be written.

When the partition-function summation can be approximated as an integral, we use our new parameter to rewrite equation (d) as:

$$z_{\text{rot}} = \sum_0^\infty (2J + 1)e^{-J(J+1)\theta_r/T} = \int_0^\infty e^{-J(J+1)\theta_r/T}(2J + 1)\, dJ$$

$$= \int_0^\infty e^{-(J^2+J)\theta_r/T} \cdot d(J^2 + J).$$

For a given substance θ_r is a constant. For a given substance at a given temperature, θ_r/T is thus a *constant* that can be moved freely across differential and integral operators. Symbolizing the constant θ_r/T as α, and the variable $(J^2 + J)$ as y, we see that the integral has the simple form

$$\int e^{-\alpha y}\, dy = -\frac{1}{\alpha}\int e^{-\alpha y}\, d(-\alpha y) = -\frac{1}{\alpha} e^{-\alpha y}.$$

Therefore

$$z_{rot} = \frac{-T}{\theta_r}\left[e^{-(J^2+J)\theta_r/T}\right]_{J=0}^{J=\infty} = \frac{-T}{\theta_r}\left[e^{-\infty} - e^{-0}\right] = \frac{-T}{\theta_r}[0 - 1] = \frac{T}{\theta_r}.$$

This unexpectedly simple result we can also write as

$$z_{rot} = \frac{T}{\theta_r} = \frac{8\pi^2 IkT}{h^2}.$$

But this preliminary finding requires two very important qualifications.

First qualification: In obtaining the above expression for z_{rot}, we approximated a summation by an integral. This approximation is acceptable at any temperature for which the spacing of the rotational energy levels is minute compared to the thermal energy kT—for under such conditions the energy levels *do* simulate a continuum we can treat by integration. Now the spacing of the energy levels is clearly a function of θ_r, for consider that

$$\epsilon_J = J(J + 1)\frac{h^2}{8\pi^2 I} = J(J + 1)k\theta_r.$$

Thus the larger θ_r, the greater will be the energy spacing between successive rotational levels—and the higher will be the temperature required to make kT much greater than the energy spacings. At all temperatures for which $\theta_r/T \le 0.01$ for the species of rotor concerned, the integration is permissible and the approximation excellent. Even when $\theta_r/T = 0.1$, the resultant error in z_{rot} is only about 0.5%. But the approximation fails badly for still higher values of the ratio θ_r/T, i.e., whenever T is too small and/or θ_r too large. For what diatomic gases may we expect conspicuously large values of θ_r? Defined as $h^2/8\pi^2 Ik$, θ_r will be conspicuously large whenever the moment of inertia (I) is conspicuously small. The smallest values of I and, hence, the largest values of θ_r are found in diatomic gases containing hydrogen. At and below room temperature, the value of θ_r/T for hydrogen itself is so very large that we must certainly forego the use of integration in evaluating the partition function.† For diatomic gases

---

† We are all the more disposed to exclude molecular hydrogen because its behavior is complicated by a particularly subtle species of (*ortho-para*) isomerism. In itself, the inapplicability of integration presents no great difficulty. For, precisely in those circumstances in which the partition-function summation must be evaluated *as* a summation, the series is always rapidly enough convergent to be easily evaluated.

containing *one* hydrogen atom, $\theta_r/T < 0.1$ at temperatures down to the boiling points of the respective gases, and integration is at least marginally acceptable at all temperatures not too far below room temperature. For diatomic gases containing *no* hydrogen, $\theta_r/T < 0.01$ at temperatures down to the boiling points of the respective gases, and integration yields excellent values of z_{rot} at *all* temperatures at which these species can be handled *as gases*. Subject only to the restrictions just noted, the replacement of the partition-function summation by an integral is thus seen to be very generally permissible—and is always permissible in the cases with which we shall be concerned.

Second qualification: From the equation $z_{rot} = T/\theta_r$, we obtain excellent values of z_{rot} for heteronuclear diatomic gases like CO, but for homonuclear diatomic gases like O_2 or N_2 we obtain values of z_{rot} just twice the correct value. This defect we make good by rewriting our equation as $z_{rot} = T/\sigma\theta_r$, where the "symmetry factor" (σ) equals *one* for heteronuclear molecules and *two* for the more symmetric homonuclear molecules. In classical terms, one may seek to rationalize this insertion by relating it to the insertion of the factor $1/N!$ in the translational partition function of an assembly. The latter insertion was justified by pointing out that, in an assembly of nonlocalized units, many imaginable "states" of the assembly are wholly indistinguishable even in principle. One may then seek to justify the insertion of σ by pointing out that, when a homonuclear molecule is rotated through 180°, its final position is indistinguishable from its initial position. This is not the case when a heteronuclear molecule is similarly rotated: the two positions are readily distinguishable in principle. Thus each pair of positions separated by 180° represents *two* distinguishable positions for a heteronuclear molecule, but only *one* position for the more symmetric homonuclear molecule. And so one may think to have justified the insertion of a symmetry factor which makes z_{rot} for a homonuclear molecule just half as great as z_{rot} for an otherwise identical heteronuclear molecule. Though not without plausibility, this style of argument seems a little slippery. An alternative argument has rather less immediate appeal—because it draws on quantum mechanics for a theorem we cannot demonstrate here—but has the great virtue of a solid foundation. We then set out from a quantum-mechanical proposition asserting that, while all the quantum states symbolized by integral values of J are simultaneously accessible to a heteronuclear rotor, a symmetric homonuclear rotor can assume only those states symbolized by *either* even values of J *or* odd values of J, *but not both*. Now whenever we replace the partition-function summation by an integral, in effect we assume that (i) the summation extends over *many* values of J, and (ii) the contributions to z_{rot} made by terms with adjacent values of J differ only *infinitesimally* from each other. Hence, when we sum over even values of J only, or over odd values of J only, the value of z_{rot} so obtained must be just *half* as great as when, other things being equal, we sum over *all* integral values of J. But this is *precisely* the effect we have then quite properly allowed for by inserting a symmetry factor σ, with value of unity for heteronuclear molecules and value of two for homonuclear molecules.

We are thus led to write for the partition function of a diatomic molecule:

$$z_{rot} = \frac{T}{\sigma\theta_r} = \frac{8\pi^2 I k T}{\sigma h^2}. \tag{23}$$

Since rotation is an internal degree of freedom, we have also

$$Z_{rot} = (z_{rot})^N = \left[\frac{8\pi^2 I k T}{\sigma h^2}\right]^N.$$

We can now easily determine the rotational contributions to the thermodynamic functions. On the strength of equation (18), the thermodynamic energy is given as

$$E_{rot} = kT^2 \left[\frac{d \ln Z_{rot}}{dT}\right]_V = kT^2 \frac{d}{dT} \ln \left[\frac{8\pi^2 I k T}{\sigma h^2}\right]^N$$

$$= NkT^2 \left[\frac{d}{dT} \ln T + \frac{d}{dT} \ln \frac{8\pi^2 I k}{\sigma h^2}\right].$$

Setting equal to zero the derivative of the last logarithmic term on the right, which involves only a collection of constants, we find:

$$E_{rot} = NkT^2 \frac{d}{dT} \ln T = NkT^2 \cdot \frac{1}{T} = NkT,$$

or, for one mole,

$$\bar{E}_{rot} = RT.$$

This implies that, as in the earlier case of the translational contribution, the rotational contribution to the total energy is wholly independent of the identity of the (diatomic) molecule concerned.

Turning then to the rotational contribution to the heat capacity, we write

$$(C_V)_{rot} = \left[\frac{d\bar{E}_{rot}}{dT}\right]_V = \frac{d}{dT} RT = R.$$

The rotational contribution to the heat capacity appears independent alike of the identity of the (diatomic) gas and the temperature concerned. However, this finding becomes unreliable at low temperature when, as earlier remarked, we can no longer rely on the integration used in deriving equation (23). And indeed when the partition function is evaluated *as* a summation, we duly find that $(C_V)_{rot} \rightarrow 0$ as $T \rightarrow 0$.

For the rotational contribution to the entropy, substitution in equation (19) yields

$$S_{rot} = \frac{E_{rot}}{T} + k \ln Z_{rot} = \frac{NkT}{T} + k \ln \left[\frac{8\pi^2 IkT}{\sigma h^2} \right]^N$$

$$= Nk \left[1 + \ln \frac{IT}{\sigma} + \ln \frac{8\pi^2 k}{h^2} \right].$$

On substituting the appropriate numerical values for all the indicated constants, we find for one mole of gas

$$\bar{S}_{rot} = R \ln IT/\sigma + 177.68. \tag{24}$$

No difficulty now attends determination of the rotational contribution to the Helmholtz free energy $A(\equiv E - TS)$. Observe that, unlike the expressions for E and C_V, the equation for S (and that for A) include the species-specific terms σ and I. The chemical formula of the diatomic gas concerned determines whether we should set σ equal to 1 or to 2, and spectroscopic data suffice to establish the value of I appropriate in each instance. Thus we can readily determine the rotational contributions to E, C_V, S, and A for all diatomic gases save hydrogen, at all temperatures well removed from $0°K$.

As in the earlier treatment of translations, we close this analysis of rotations with a minimum estimate of the number of accessible rotational quantum states. Again we count only those states with energies ranging up to but not beyond the *average* energy, which the result $\bar{E}_{rot} = RT$ per mole implies to be $\epsilon = \bar{E}/N = kT$ per molecular rotor. To discover the value of J that corresponds to this mean energy, we call on the energy-quantization condition for rotations, to write

$$(J)(J + 1) \frac{h^2}{8\pi^2 I} = \epsilon = kT,$$

$$(J)(J + 1) = \frac{8\pi^2 IkT}{h^2}.$$

The function on the right is nothing but the partition function (z_{rot}) for a heteronuclear rotor. And we then suspect that $(J)(J + 1)$ must somehow represent the number of accessible rotational quantum states. This suspicion is easily substantiated.

Each rotational energy level is characterized by a degeneracy ranging from 1 when $J = 0$ up to $(2J + 1)$ for whatever value of J the last equation establishes as correspondent to the rotational energy kT. The *average* number of rotational quantum states per rotational energy level can then be established simply by averaging the first and last terms in the arithmetic series 1, 3, 5, ... $(2J + 1)$. We so obtain

$$\frac{1 + (2J + 1)}{2} = J + 1.$$

In J energy levels with a mean degeneracy of $J + 1$, the total number of rotational quantum states is of course just $(J)(J + 1)$. Even as before in the case of translations, we have thus found for rotations too that the partition function is excellently approximated by a minimum estimate of the total number of accessible quantum states. The basis for this relationship is the very same as that given before (on p. 76). And just as before, we are again in a position to offer a convincing rationalization of an entropy expression. With W now estimated as the number of ways in which N distinguishable rotors can be distributed over $(J)(J + 1)$ rotational quantum states assumed (counterfactually) to be equivalent in energy, **problem 22** invites us to use the definition $S \equiv k \ln W$ to estimate the rotational contribution to the entropy of a heteronuclear diatomic gas. One then obtains an expression almost identical with equation (24), the only difference being that in this crude estimate the numerical constant differs by about 1% from the figure that appears in equation (24).

All the foregoing arguments, all the foregoing conclusions, are just like those developed in our earlier analysis of translations. But of course rotations *do* differ from translations, and most conspicuously in our minimum estimate of the number of accessible quantum states. For rotations, that minimum estimate is represented by the product $(J)(J + 1)$, which we can easily determine by making appropriate numerical substitutions in the expression on p. 86. For CO, a representative heteronuclear diatomic molecule with $I = 14.5 \times 10^{-40}$ gm-cm^2, we find that at 298°K the product $(J)(J + 1) \simeq 10^2$. The meagre size of this number alerts us to the potential vulnerability of our replacement of the partition-function summation by an integral. And, comparing 10^2 with the 10^{30} that represents our estimate of the number of accessible translational quantum states, we correctly infer that the energy spacing between contiguous rotational states is much broader than that between contiguous translational states. This conspicuous *physical* difference is what is reflected in the conspicuous *numerical* difference between $z_{rot}(\simeq 10^2)$ and $z_{trans}(\simeq 10^{30})$.

Vibration. The simplest model for the vibration of a diatomic molecule is the so-called "harmonic oscillator." Imagine a weightless spring, with relaxed length r, which obeys Hooke's law. That is, a spring which—when stretched or compressed to some length l—resists further stretching or compression with a force proportional to $(l - r)$. The proportionality constant (κ) is known as the "force-constant" of the spring. Now imagine two masses, m_1 and m_2, placed on a frictionless horizontal surface, and

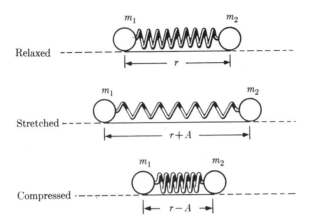

Figure 17

linked by the spring just described.   When set in oscillation around the (stationary) center of mass of the system, the two masses vibrate with some amplitude A defined, as indicated in Fig. 17, in terms of their closest approach and greatest separation.   The vibration takes place with a frequency v determined by the formula:

$$v = \frac{1}{2\pi} \sqrt{\frac{\kappa}{\mu}},$$

where again the "reduced mass" $\mu \equiv m_1 m_2 / (m_1 + m_2)$.

The frequency is independent of the amplitude, but the total energy (ϵ) associated with the oscillation *is* a function of its amplitude.   In classical physics there is no restriction on the permissible amplitudes and, to the continuum of amplitudes then possible, there corresponds a continuum of possible vibrational energies given by the formula $\epsilon = \frac{1}{2}\kappa A^2$. But in quantum physics this is no longer the case for an oscillator of microcosmic dimensions. An additional restriction then operates to reduce the permissible energies to the noncontinuous set of discrete values defined by the equation:

$$\epsilon = (v + \tfrac{1}{2})hv,$$

where the vibrational quantum number v can assume only the *integral* values 0, 1, 2, ... Actual molecular oscillators will only be approximated by this very simple expression for the vibrational quantization condition of an ideal harmonic oscillator.   However, the harmonic-oscillator approximation will be quite good enough for all the cases here at issue.

Concerned as we are primarily with *diatomic* molecules, we need say little of the very complicated vibrational patterns that occur in large molecules. Just as any linear velocity can be conceptually constructed from x-, y-, and z-components of velocity, even so a complicated vibration can be conceptually constructed from components of a limited set of "normal-mode" vibrations. How many such vibrational degrees of freedom need we consider? For N independent gaseous atoms in a box, we must make $3N$ specifications of translational quantum numbers (3 per atom present) to obtain a full specification of the state of the system. When the N atoms are united in one molecule, we suppose that $3N$ specifications are still required fully to define the state of the system, and we then speak of the molecule as having a total of $3N$ degrees of freedom. To conform with this requirement, the number of independent vibrational degrees of freedom must be assigned as indicated in the following table.

	Degrees of Freedom	
Translational motion of molecule's center of gravity	3	
Rotational motions		
Linear molecule	2	
Nonlinear molecule		3
Fundamental vibrational modes		
Linear molecule	$3N - 5$	
Nonlinear molecule		$3N - 6$
Linear molecule	$3N$	
Nonlinear molecule		$3N$

According to whether it is linear or nonlinear, a molecule will thus have either $(3N - 5)$ or $(3N - 6)$ vibrational degrees of freedom. For a diatomic molecule, *necessarily* linear with $N = 2$, we have $3N - 5 = 1$—which means that the molecule has only the one vibrational degree of freedom indicated for the model sketched in Fig. 17.

All preliminaries being now completed, we turn to evaluation of the partition function of a harmonic oscillator. If as usual we identify our reference zero of energy with the ground-state energy $\epsilon_0 \equiv 0$, we obtain $\epsilon_1 = hv$, $\epsilon_2 = 2hv$, and so on. For the vibrational partition function we will then write

$$z_{vib} = \sum e^{-\epsilon_{vib}/kT} = 1 + e^{-hv/kT} + e^{-2hv/kT} + \cdots$$

Having the form $z_{vib} = 1 + e^{-x} + e^{-2x} + e^{-3x} + \cdots$, this geometric series is easily shown to be equal to the quotient $1/(1 - e^{-x})$. For by actual long division we find

$$
1 - e^{-x} \overline{)1} \quad \frac{1 + e^{-x} + e^{-2x} + e^{-3x} + \cdots}{}
$$

$$
\underline{1 - e^{-x}}
$$

$$
+ e^{-x}
$$

$$
\underline{+ e^{-x} - e^{-2x}}
$$

$$
+ e^{-2x}
$$

$$
\underline{+ e^{-2x} - e^{-3x}}
$$

Making *no* approximation, we can therefore easily express the partition function in closed analytical form, as

$$
z_{\text{vib}} = \frac{1}{1 - e^{-hv/kT}}. \tag{25}
$$

For a representative diatomic molecule, $v \simeq 6 \times 10^{13}$ sec^{-1}, $hv/k \simeq 3000°$K, and at $298°$K

$$
z_{\text{vib}} \simeq \frac{1}{1 - e^{-3000/300}} = \frac{1}{1 - e^{-10}} \simeq 1.
$$

This implies that at room temperature only *one* vibrational state (i.e., the ground state) is ordinarily accessible to diatomic molecules—a situation quite different from that earlier found to characterize rotations, and *very* different from that for translations.

For an internal degree of freedom like vibration, equation (15) gives $Z_{\text{vib}} = (z_{\text{vib}})^N$. Determination of the vibrational contribution to the total thermodynamic energy then presents no difficulty. Since no term in the vibrational partition function is volume-dependent, we can drop the constant-volume rider on equation (18), to find

$$
E_{\text{vib}} = kT^2 \frac{d \ln Z_{\text{vib}}}{dT} = kT^2 \frac{d \ln (z_{\text{vib}})^N}{dT} = NkT^2 \frac{d \ln z_{\text{vib}}}{dT}
$$

$$
= NkT^2 \frac{d}{dT} \ln \frac{1}{1 - e^{-hv/kT}} = -NkT^2 \frac{d}{dT} \ln (1 - e^{-hv/kT})
$$

$$
= -NkT^2 \frac{1}{1 - e^{-hv/kT}} (-e^{-hv/kT}) \left(\frac{-hv}{k} \right) \left(\frac{-1}{T^2} \right)
$$

$$
= +NkT^2 \left(\frac{hv}{kT^2} \right) \frac{e^{-hv/kT}}{1 - e^{-hv/kT}}.
$$

For one mole of units, therefore,

$$
\bar{E}_{\text{vib}} = R \left(\frac{hv}{k} \right) \frac{1}{e^{hv/kT} - 1}. \tag{e}
$$

This differs greatly from the simple expressions $\bar{E}_{trans} = \frac{3}{2}RT$ and $\bar{E}_{rot} = RT$. But recall that these earlier results were derived from partition functions obtained with the aid of an integration that, in effect, presumes a high-temperature limit. And in that limit we do indeed find a similarly uncomplicated expression for $\bar{E}_{vib}$ also. With $e^x = 1 + x + x^2/2! + x^3/3! + \cdots$, in the limit of high temperatures, when $(hv/kT) \ll 1$,

$$\bar{E}_{vib} = R\left(\frac{hv}{k}\right) \frac{1}{1 + \frac{hv}{kT} + \cdots - 1} = R\left(\frac{hv}{k}\right) \frac{1}{\frac{hv}{kT}} = RT.$$

This similarity notwithstanding, the vibrational case is sharply distinguished from rotations and translations. For those degrees of freedom, involving quantum states that are rather narrowly spaced in energy, room temperature represents a "high" temperature. With the comparatively wide energy spacing of the vibrational quantum states, on the other hand, room temperature there figures as a very "low" temperature. Already demonstrated by our finding that, in a representative case, $z_{vib} \simeq 1$ at room temperature, this conclusion can be further documented by a calculation of $\bar{E}_{vib}$ for the same case. Again taking $hv/k = 3000°K$, we find from equation (e) that at $300°K$

$$\bar{E}_{vib} = R(3000) \frac{1}{e^{10} - 1} \simeq R \frac{3000}{21000} = \frac{1}{7} R.$$

Remote from the high-temperature limit in which we would have found $300R$, this very small value confirms our earlier inference that, at room temperature, the vast majority of these units must remain in their ground state.

What would we have found had we chosen to set our reference zero of energy *not* at the ground state but, rather, at that level (*vide infra*) from which vibrational energies are measured by the quantization condition, $\epsilon = (v + \frac{1}{2})hv$? On that scale the energies of the quantum states are $\epsilon_0 = \frac{1}{2}hv$, $\epsilon_1 = \frac{3}{2}hv$, $\epsilon_2 = \frac{5}{2}hv$, and so on. Symbolizing by z' the partition function now to be calculated relative to our new reference zero of energy, we write

$$z'_{vib} = \sum e^{-\epsilon'_{vib}/kT} = e^{-hv/2kT} + e^{-3hv/2kT} + e^{-5hv/2kT} + \cdots$$
$$= e^{-hv/2kT}(1 + e^{-hv/kT} + e^{-2hv/kT} + \cdots).$$

The summation in parenthesis is exactly the same as that earlier found to correspond to the partition function z_{vib}. Hence

$$z'_{vib} = z_{vib} \cdot e^{-hv/2kT}.$$

Following the same line that before led us from equation (18) to equation (e), we may now derive an expression for the energy ($\bar{E}'_{vib}$) corresponding to the partition function z'_{vib}. This derivation is the subject of **problem 24(a)**, to which the answer is

$$\bar{E}'_{vib} = \bar{E}_{vib} + N(\tfrac{1}{2}h\nu)$$

where $\bar{E}_{vib}$ is exactly as given in equation (e).

This is an eminently satisfactory result. Relative to our second choice of a reference zero for energy, the quantity $N(\tfrac{1}{2}h\nu)$ is simply the energy $(\bar{E}^0_{vib})$ possessed by the N units when they *all* stand in the ground state. And the previously derived expression for $\bar{E}_{vib}$ clearly represents the *surplus* energy possessed by the units when they occupy states *other than* the ground state. Thus we have every reason to expect what we have found: namely, $\bar{E}'_{vib} = \bar{E}_{vib} + \bar{E}^0_{vib}$. For *any* reference zero of energy relative to which $\bar{E}^0_{vib}$ symbolizes the energy of a mole of units in the ground state, the energy $\bar{E}'_{vib}$ calculated from the partition function z'_{vib} will thus be linked with $\bar{E}_{vib}$ by the equation

$$\bar{E}'_{vib} - \bar{E}^0_{vib} = \bar{E}_{vib} = R\left(\frac{h\nu}{k}\right)\frac{1}{e^{h\nu/kT} - 1}.$$

Since heat capacities express only *differences* of energy, we may be sure that the expression for the vibrational contribution to the heat capacity will emerge alike from any choice of a reference zero for energy. Returning to the expression for $\bar{E}_{vib}$ given in equation (e), we write

$$C_V = \left[\frac{d\bar{E}}{dT}\right]_V = \frac{d}{dT}\left[R\left(\frac{h\nu}{k}\right)\frac{1}{e^{h\nu/kT} - 1}\right] = R\left(\frac{h\nu}{k}\right)\frac{d}{dT}\left[\frac{1}{e^{h\nu/kT} - 1}\right]$$

$$= R\left(\frac{h\nu}{k}\right)\left[\frac{-1}{(e^{h\nu/kT} - 1)^2}\,(e^{h\nu/kT})\left(\frac{-h\nu}{kT^2}\right)\right].$$

Collection of terms then yields

$$(C_V)_{vib} = R\left(\frac{h\nu}{kT}\right)^2\frac{e^{h\nu/kT}}{(e^{h\nu/kT} - 1)^2}.$$

In this equation the characteristic frequency ν is the only term that reflects the identity of the oscillators concerned. Just as we earlier defined a rotational parameter θ_r, so here we define a species-specific vibrational parameter $\theta_v \equiv h\nu/k$. In terms of θ_v, we can rewrite the last equation more compactly, as

$$(C_V)_{vib} = R\left[\left(\frac{\theta_v}{T}\right)^2\frac{e^{\theta_v/T}}{(e^{\theta_v/T} - 1)^2}\right]. \tag{26}$$

We defer for the time being a full account of how this expression, first derived by A. Einstein, yielded one of the earliest major triumphs of the quantum viewpoint.

Turning then to the entropy, we call on equation (19) to find

$$S_{vib} = \frac{E_{vib}}{T} + k \ln Z_{vib}$$

$$= Nk \left(\frac{hv}{kT}\right) \frac{1}{e^{hv/kT} - 1} + k \ln (z_{vib})^N$$

$$= Nk \left(\frac{hv}{kT}\right) \frac{1}{e^{hv/kT} - 1} + Nk \ln \frac{1}{1 - e^{-hv/kT}}$$

$$= Nk \left[\frac{\theta_v}{T} \cdot \frac{1}{e^{\theta_v/T} - 1} - \ln (1 - e^{-\theta_v/T})\right].$$

No difficulty now attends derivation of an expression for the Helmholtz free energy $A(\equiv E - TS)$. Since S (like C_V) is independent of the reference zero chosen for calculation of the partition function, that choice obviously must affect A precisely as it does E.

Electronic excitation. Just as the hydrogen atom can exist in excited states, with energies higher than that of the ground state, so also molecules can exist with electronic arrangements to which correspond energies higher than that of the electronic ground state. The energy spacing of these states varies irregularly, so that no general expression for the electronic partition function is possible. However, this causes no problem in ordinary practice. With only rare exceptions, including a very few common species like NO (see **problem 25**), all the excited electronic states are associated with energies *much* greater than that of the electronic ground state. So much so that the excited states remain completely unpopulated at temperatures up to several thousand degrees Kelvin. But, with all units thus confined to the ground state, the evaluation of the partition-function summation reduces to a calculation of no more than its first (ground-state) term. Note well that this does *not* mean that in all such cases the partition function $z_{elec} = 1$. For observe:

$$z_{elec} = \sum \omega_y e^{-\epsilon_y/kT} = \omega_0 e^{-\epsilon_0/kT} + \omega_1 e^{-\epsilon_1/kT} + \omega_2 e^{-\epsilon_2/kT} + \cdots$$

Setting $\epsilon_0 \equiv 0$, and assuming that ϵ_1, ϵ_2, etc. are much greater than kT (i.e., all excited states energetically remote from the ground state), we then find:

$$z_{elec} = \omega_0 e^{-0} + \omega_1(0) + \omega_2(0) + \cdots = \omega_0.$$

Thus even when all the excited states remain wholly untenanted, degeneracy of the electronic ground state must always yield a partition function $z_{elec} > 1$. When we come to consider how equilibrium constants can be derived from spectroscopic data, we will find an important role for this limiting relation, $z_{elec} = \omega_0$.

The last paragraph involves one conspicuous oversimplification. In evaluating partition functions, always heretofore we have followed the dictates of convenience in setting the reference zero of energy to coincide with the ground state, i.e., $\epsilon_0 \equiv 0$. So long as we are concerned with only *one* degree of freedom of only *one* species of unit, we are certainly free to set the energy reference zero wherever we please. However, on many occasions (e.g., in a forthcoming section on chemical equilibrium) attention shifts to the *relation* among thermodynamic magnitudes calculated from the respective partition functions for *different* species of units. Clearly the meaningfulness of any such interrelationship will be secure only if all the partition-function calculations are based on the *same* reference zero of energy.

When all units in an assembly of gaseous atoms occupy the translational ground state (as presumably they would in a gas at $0°K$), the atoms are effectively at rest† with respect to their container. Let us now set up, in coincidence with the translational ground state of gaseous atoms, what we will hereafter call the SSA (stationary-separated-atom) reference zero of energy. The SSA reference zero obviously coincides also with the translational ground state of gaseous molecules. And the same SSA reference zero coincides further with the ground states of the rotational degrees of freedom open to molecules but not to atoms. For when $J = 0$ the units possess no angular momentum, and so may be regarded as at rest with respect to their container. Thus our habitual identification of the reference zero of energy with the ground state can now be seen to assure us of something more than convenience in calculation. That identification assures a perfectly consistent use of the *same* SSA reference zero for *all* translational and *all* rotational degrees of freedom of *all* species of gaseous units. However, in the vibrational and electronic degrees of freedom we encounter a very different situation.

A molecule exists as a stable entity precisely *because* its potential energy is less than that possessed by the separate atoms of which it is composed. That is, the atoms stay together because they are trapped inside a potential-energy well like those shown in Fig. 18. Moreover, it is relative to the bottom of the potential well that the vibrational energies are expressed by the quantization condition $\epsilon = (v + \frac{1}{2})h\nu$. Thus when we calculated the vibrational partition function relative to the ground-state energy $\epsilon_0 \equiv 0$, we certainly were *not* operating on the SSA scale of energies. For the vibrational ground state may now be seen to fall at some value $-\epsilon_d$ below the SSA reference level. Still worse, the margin of difference between the SSA zero and the energy of the vibrational ground state is not only *large*, but also highly *variable*. Apart from the variation of ϵ_d from one electronic state to

---

† We say *effectively* at rest because, though even at $0°K$ the atoms possess some translational "zero-point energy," this is negligible—amounting to only about 10^{-12} times the average translational energy of the atoms at room temperature.

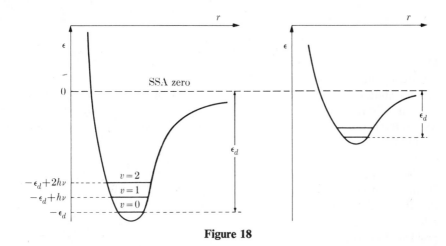

Figure 18

another of the same molecular species, the value of ϵ_d varies markedly from one species of molecule to another—as indicated in Fig. 18.

This variation is hardly surprising when we consider what is represented by the quantity ϵ_d. For a diatomic molecule, ϵ_d is simply the energy of dissociation at $0°K$. That is, the energy required to produce atoms, at rest when "separated to infinity," from a molecule originally at rest and in its vibrational ground state. Of course, we can make no actual measurement at $0°K$. But from spectroscopic measurements, at readily accessible temperatures, we can easily calculate the value of ϵ_d. And now we can state our problem. Given a determination of ϵ_d for each particular species in question, how can we convert all our evaluations of vibrational and electronic partition functions to the same SSA reference zero used in evaluating rotational and translational partition functions? Once so stated, the problem is readily solved.

Ordinarily we need consider nothing but the electronic ground state of the molecule concerned, and when (as is often the case) $\omega_0 = 1$ we may omit z_{elec} altogether. We shall therefore attach to z_{vib} the entire correction term required to place all our calculations on the SSA standard. Relative to this reference zero, the ground state of a harmonic oscillator will be associated with energy $-\epsilon_d$, and the excited states with energies $(-\epsilon_d + h\nu)$, $(-\epsilon_d + 2h\nu)$, and so on. Symbolizing by $\tilde{z}_{vib}$ the vibrational partition function calculated on the SSA basis, we write:

$$\tilde{z}_{vib} = \sum e^{-\epsilon_q/kT} = e^{-(-\epsilon_d)/kT} + e^{-(-\epsilon_d + h\nu)/kT} + e^{-(-\epsilon_d + 2h\nu)/kT} + \cdots$$

$$= e^{+\epsilon_d/kT}(1 + e^{-h\nu/kT} + e^{-2h\nu/kT} + \cdots).$$

The summation in parentheses is seen to be identical with that we found (p. 89) to represent the partition function (z_{vib}) calculated on the basis

$\epsilon_0 \equiv 0$. Hence,

$$\tilde{z}_{vib} = z_{vib} \cdot e^{+\epsilon_d/kT}. \tag{27}$$

Using the value of ϵ_d appropriate to the species of oscillator in question, we are to multiply, by $e^{+\epsilon_d/kT}$, the value of z_{vib} calculated relative to the ground state $\epsilon_0 \equiv 0$. And, so doing, we easily attain what we set out to find: a vibrational partition function ($\tilde{z}_{vib}$) consistently reduced to the SSA reference zero. Thus *all* partition-function calculations for *all* species of units can indeed be based on one and the same SSA reference zero of energy.

Where before we wrote $Z_{vib} = (z_{vib})^N$, now we will write $\tilde{Z}_{vib} = (\tilde{z}_{vib})^N$. By insertion of $\tilde{Z}_{vib}$ in the appropriate equations, we can then derive for all the thermodynamic parameters values consistently reduced to the SSA reference standard. These quite routine derivations form the substance of **problem 24**—solution of which makes it quite clear why the shift to the SSA reference zero introduces a new additive term into the expression for A_{vib} but not into that for S_{vib}.

Polyatomic gases. While we attempt no general treatment of polyatomic gases here, it may be worth noting briefly how they can be brought under the scheme of analysis we have developed.

To evaluate the translational partition function for a polyatomic gas, we can proceed exactly as we did for a monatomic gas: we need only introduce into equation (21) the mass of the polyatomic molecule concerned.

In the evaluation of the vibrational partition function for a polyatomic gas, we also encounter no fundamental new complications. To be sure— unlike the diatomic molecule with its single vibrational degree of freedom— a polyatomic molecule offers scope for multiple vibrational degrees of freedom. Each such vibration has its own (spectroscopically assignable) characteristic frequency and, hence, its own characteristic value of $\theta_v(= h\nu/k)$. In the absence of serious coupling between rotational and vibrational degrees of freedom, we can separately evaluate a partition function ($z_{vib\,1}$, $z_{vib\,2}$, $z_{vib\,3}$, ...) for *each* vibration—just as we do when, in a diatomic molecule, only *one* vibration is possible. And then we establish the complete vibrational partition function for the polyatomic molecule simply by multiplying to obtain: $z_{vib} = z_{vib\,1} \cdot z_{vib\,2} \cdot z_{vib\,3} \cdots$.

To evaluate the rotational partition function for a *linear* polyatomic molecule, we can proceed exactly as in making that evaluation for a diatomic molecule. As in the diatomic case, the linear polyatomic molecule still has only *two* rotational degrees of freedom and a single moment of inertia (I). Hence, as before, we can use equation (23):

$$z_{rot} = \frac{8\pi^2 IkT}{\sigma h^2}.$$

For *nonlinear* polyatomic molecules, on the other hand, there will be three principal moments of inertia (I_1, I_2, I_3) associated with the *three* degrees of freedom represented by rotations of the molecule around three mutually perpendicular axes intersecting at its center of mass. The more elaborate formula then required is here simply *announced*:

$$z_{rot} = \frac{8\pi^2}{\sigma h^3} (I_1 \cdot I_2 \cdot I_3)^{1/2} (2\pi kT)^{3/2}.$$

The symmetry factor σ can again be established simply by counting the number of different positions in which the molecule concerned presents an identical appearance: e.g., for benzene, a planar hexagonal structure, $\sigma = 12$.

Though somewhat tedious, evaluation of these partition functions presents no new problem in principle. However, when a polyatomic molecule is nonrigid—as larger molecules often are—very nasty problems arise from the possible occurrence of large-scale motions in which one part of the molecule rotates relative to another part. When such "internal rotation" is strongly hindered, it can be treated adequately as a vibration; i.e., the two parts of the molecule undergo a rotary oscillation relative to one another. And when such internal rotation is only slightly hindered, it can be treated adequately as a rotation. But, alas, many actual cases fall between these simple extremes—in a region where proper evaluation of partition functions presents grave (but not insuperable) difficulties. Beyond noting that such problems exist, we need not further concern ourselves with them.

Applications 4

Even as it was being developed, our statistical mechanics has already been somewhat applied, e.g., in derivation of the Sackur-Tetrode equation and the Maxwell-Boltzmann speed-distribution law. But with the development phase now behind us, in the present chapter the sphere of application becomes our primary concern.

HEAT CAPACITIES

We give first consideration to heat capacities because, of all the thermodynamic parameters that may be calculated from the partition function, heat capacities are most directly given by experimental measurements with which we may then compare our calculated results. Beginning with an assembly that has only vibrational degrees of freedom, we here encounter the only application of our statistical mechanics to a condensed phase.

The Einstein solid. Consider an isotropic crystalline element consisting of N atoms. Let us suppose that the thermal vibrations of these units in the lattice are (i) well approximated as harmonic oscillations all having the *same characteristic frequency*, v; (ii) substantially independent from one unit to another, so that the vibrational state of the first does not appreciably predetermine the vibrational state of the second; and (iii) substantially independent from one vibrational degree of freedom to another of the same atom, so that the state of the atom's vibration in (say) the x-direction does not appreciably predetermine its vibrational states in the y- and z-directions. To the extent that these rather daring assumptions are valid, we may regard the crystal as an assembly of $3N$ one-dimensional harmonic oscillators—three per atom present. This is of course a gross oversimplification: we can hardly imagine that the vibrations of neighboring atoms represent genuinely *independent* degrees of freedom. However, the same conclusion emerges from a more acceptable analysis. If the crystal is conceived as a single non-linear "molecule" containing N atoms, the table on p. 89 implies the presence of $3N - 6$ independent normal-mode vibrations. For very large N this is equivalent to saying that the crystal has $3N$ independent vibrational degrees of freedom—exactly in line with our earlier conclusion.

Now for one mole of the crystalline element the number of atoms is Avogadro's number (N), so that one mole will represent an assembly of 3N one-dimensional oscillators. From equation (26), which refers to an assembly of just N one-dimensional oscillators, we can at once draw the following expression for the molar heat capacity of the crystalline element:

$$C_V = 3R \left[\left(\frac{\theta_v}{T} \right)^2 \frac{e^{\theta_v/T}}{(e^{\theta_v/T} - 1)^2} \right]. \qquad (a)$$

What does this formula entail? Recalling that the function e^x is equivalent to the sum of the infinite series, $1 + x + x^2/2! + x^3/3! + \cdots$, we can readily determine the import of equation (a) in two limiting cases.

As the first limiting case, let us consider the situation at very high temperature, when $T \gg \theta_v$ and $\theta_v/T \ll 1$. On expanding both the exponential terms in equation (a) we find:

$$C_V = 3R \left(\frac{\theta_v}{T} \right)^2 \frac{1 + \theta_v/T + (1/2!)(\theta_v/T)^2 + (1/3!)(\theta_v/T)^3 + \cdots}{[(1 + \theta_v/T + (1/2!)(\theta_v/T)^2 + (1/3!)(\theta_v/T)^3 + \cdots) - 1]^2}.$$

After cancelling the 1's in the denominator, we need retain there only the θ_v/T term since, with $1 \gg \theta_v/T$, the higher powers of θ_v/T will be entirely negligible. For the same reason, we can drop from the numerator everything but the 1. All that now remains is

$$C_V = 3R(\theta_v/T)^2 \frac{1}{(\theta_v/T)^2} = 3R \simeq 6 \text{ cal/mole-°K.}$$

Thus we learn that the Einstein function—the multiplier in brackets in equation (a)—assumes a limiting value of 1 at very high temperatures. The molar heat capacity then approaches a constant limiting value which is completely independent of the vibrational frequency (and the value of θ_v) characteristic of the species of oscillator concerned. But this result of theoretical analysis represents a substantial achievement! For we have just *derived* Dulong and Petit's law†—as a *limiting law* applicable to *all* isotropic atomic crystals at sufficiently high temperatures. And we have in this fashion done rather more than merely rationalize the existence of

---

† Discovered in 1819, as a purely empirical relation applicable at room temperature, this law was generally written in the obviously equivalent form:

atomic weight $\times$ specific heat $\simeq 6$.

Dulong and Petit's law—which was adequately interpretable in classical terms—since we have also laid a basis for rationalizing the many puzzling *failures* of Dulong and Petit's law.   Classical statistical mechanics long struggled in vain to master the challenge posed by these conspicuous shortcomings.   But, as Einstein was the first to show, this challenge can be triumphantly accepted by precisely the kind of quantum statistical mechanics we have developed.   That is, we can now explain why Dulong and Petit's law fails for *some* substances even well above room temperature, and for *all* substances well below room temperature.   How does this come about?

As the second limiting case, let us consider the situation at very low temperature.   When $T \to 0$, we will have $T \ll \theta_v$, $\theta_v/T \to \infty$, and $T/\theta_v \to 0$. Returning to equation (a) we write

$$C_V = 3R \left(\frac{\theta_v}{T}\right)^2 \frac{e^{\theta_v/T}}{(e^{\theta_v/T} - 1)^2} \simeq \frac{3R}{(T/\theta_v)^2} \frac{e^{\theta_v/T}}{(e^{\theta_v/T})^2} = \frac{3R}{(T/\theta_v)^2} \frac{1}{e^{\theta_v/T}}$$

$$= \frac{3R}{(T/\theta_v)^2} \cdot \frac{1}{1 + \theta_v/T + (1/2!)(\theta_v/T)^2 + (1/3!)(\theta_v/T)^3 + \cdots}$$

$$= \frac{3R}{(T/\theta_v)^2 + T/\theta_v + 1/2! + (1/3!)(\theta_v/T) + \cdots}.$$

As $T \to 0$, the first two terms in the denominator also approach zero, but the fourth and all subsequent terms approach infinity.   Hence, as $T \to 0$, $C_V \to 0$.   Thus we readily understand how it happens that, in the limit of $T \to 0$, the molar heat capacities of *all* elements will slide off from the Dulong-Petit 6 toward zero, as shown in Fig. 19.   And we understand too why *some* elements may fall seriously short of the Dulong-Petit value even at comparatively high temperatures.   For an element with a notably large value of θ_v, even at temperatures well above 0°K the ratio θ_v/T remains large, and so entails that C_V shall remain very small.   Where may we expect to find

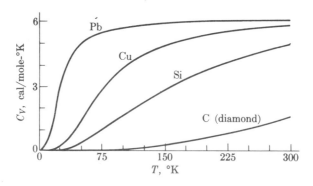

Figure 19

such behavior? For an element with an unusually large value of $\theta_v(= hv/k)$, the frequency v must be unusually high. By a formula of classical physics (see p. 88), v is in turn related to two other parameters characteristic of the species of harmonic oscillator concerned. The formula $v = (1/2\pi)\sqrt{\kappa/\mu}$ indicates that conspicuously large values of v will occur when the oscillator's force constant (κ) is particularly large and/or when its reduced mass (μ) is particularly small. Hence we are led to expect unusually high values for v (and θ_v) whenever,we are dealing with elements of low atomic weight—and the more so the harder the crystals they form. Given Einstein's formula, we have thus been led to *predict* the failures of Dulong and Petit's law in precisely those instances in which it is, in fact, known to fail. That is, the lighter elements—and particularly those that form very hard crystals—attain the limiting heat capacity of 6 only at quite high temperatures. Observe in Fig. 19 the very sharp contrast in the behavior of lead and diamond.

Everything hinges on the relative magnitudes of hv and kT, reflected in the ratio θ_v/T. When hv is large relative to kT, the partition function $z(\equiv \sum e^{-\epsilon_q/kT})$ is effectively reduced to 1—because all except the first term in the summation will be negligibly small. Essentially all the units must then fall in the ground state, and the vibrational degree of freedom remains "*unexcited*" at all temperatures for which $kT \ll hv$. Throughout this range of temperature C_V remains close to zero, simply because a temperature rise of 1° is insufficient to promote any appreciable number of units even to the first vibrational state. We have here a typical quantum phenomenon wholly unanticipated, and unanticipatable, on classical grounds. What then of the opposite extreme, reached at temperatures sufficiently high that $kT \gg hv$? In this range of temperatures the energy differences between the various vibrational states are small compared to the available (thermal) excitation energies. To the extent that the vibrational states then simulate an energy-continuum, the vibrations will be well approximated as "*classical*" degrees of freedom—with a corresponding classically derivable value of $3R$ for the molar heat capacity at constant volume. Thus (here as everywhere else) the quantum-mechanical view of the situation shades over into the classical view as a limiting case.

What lies between the limiting case of a vibrational degree of freedom adequately describable in classical terms, and the other extreme of a completely unexcited degree of freedom? Here we encounter a transition region of *partial* excitation explicable only in quantum-mechanical terms, but readily explicable in those terms. We have seen that the Einstein function $[= (\theta_v/T)^2 \cdot (e^{\theta_v/T}) \cdot (e^{\theta_v/T} - 1)^{-2}]$ approaches zero when $T \ll \theta_v$, and approaches unity when $T \gg \theta_v$. Systematic computation of the value of the function for intermediate values of θ_v/T yields the results graphed in Fig. 20. In place of θ_v/T, its reciprocal T/θ_v has been chosen as abscissa in this figure—so that the temperature may be shown to increase in the

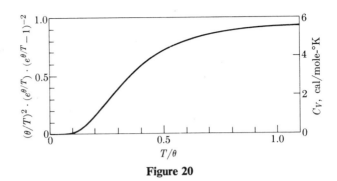

Figure 20

familiar direction, from left to right. The left-hand ordinate displays the possible values of the Einstein function; the right-hand ordinate displays the corresponding values of C_V calculated from equation (a). This graph brings a remarkable achievement within our reach. For any specified element, once given a measurement of C_V at some *one* temperature at which $0 \ll C_V \ll 6$, we can at once go on to calculate the values of C_V at *all* temperatures! How? From the curve in Fig. 20 we can read the value of T/θ_v corresponding to the value of C_V measured at a known temperature T. Possessing figures for both T and T/θ_v, we can then easily establish θ_v. But once we know θ_v for the element at issue, all the T/θ_v values along the abscissa of Fig. 20 can be converted to the corresponding figures for T, to yield a complete plot of C_V as a function of T. Figure 21 displays the excellent agreement of experimental data for diamond with the theoretical curve calculated in this fashion by Einstein.

In Fig. 22 we display a more searching comparison of the theoretical calculation with modern experimental data for lead at low temperature. Once again the agreement is very good, but at the lowest temperatures we note a systematic divergence of the theoretical curve from the experimental

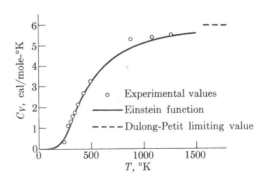

Figure 21

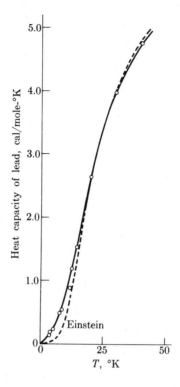

Figure 22

data. This divergence is typical and signifies, we now know, the margin by which actual crystals fail to conform to the three assumptions made in setting up the theory. Particularly vulnerable in practice is the assumption that all the 3N vibrations are characterizable by the *same* frequency v. Fortunately, this assumption becomes quite unnecessary in the refinement of Einstein's theory devised by Debye, who worked with a species-specific parameter $\theta_D \equiv hv_m/k$, where v_m is the *maximum* vibrational frequency characteristic of the substance. In terms solely of T and θ_D, one then obtains a more complex expression for C_V which yields values in excellent agreement with experimental data even at very low temperatures.†

One last obvious but impressive consequence of our analyses remains to be noted. As is evident from Fig. 19, different elements—characterized by quite different values of θ_v (or θ_D)—must yield quite different plots of C_V vs T. The rise toward the limiting value of C_V is rapid when θ is small; slow when θ is large. Comparing two different elements at the same low temperature, we are then likely to observe two *different* values of C_V. But

---

† At these temperatures the measured heat capacities often fit a T^3-law entirely concordant with a very simple limiting form of the Debye expression, namely: $C_V = 465 \, (T/\theta_D)^3$ cal/mole-°K.

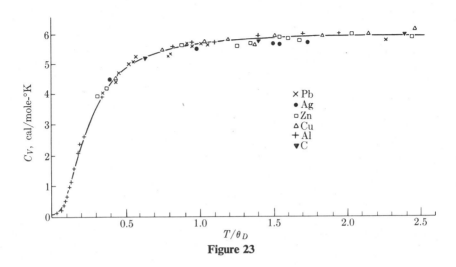

Figure 23

suppose we compare one element, at some temperature T_1, with a second element at a temperature T_2 so chosen that $T_2/\theta_2 = T_1/\theta_1$. Equation (a) then predicts, and experience confirms, that the two elements will have the *same* value of C_V. When the ratio T/θ is the same for each element, they are we say in "corresponding states." And when in corresponding states ALL elements should manifest the SAME C_V. To substantiate this conclusion, we need only recall that, for any given element, a single measurement of C_V at a known temperature suffices to determine the characteristic parameter θ. We can then easily plot, as a function of T/θ for that element, all the values of its C_V measured at other temperatures. Our conclusion leads us to expect that for ALL elements all the points will fall on identically the same curve. Schrödinger's plot of actual empirical data, displayed in Fig. 23, handsomely confirms this bold prediction. Comparing Fig. 23 with Fig. 19, who can fail to be impressed by the enormous unifying power of our theory?

Diatomic gases. Let us assume weak coupling among the one vibrational, two rotational, and three translational degrees of freedom possessed by a diatomic molecule. Let us further assume that the vibrational oscillation is truly harmonic, that the molecular rotor is truly rigid, etc. Actually, no real diatomic molecule quite meets these exacting conditions, and in refined calculations small correction terms must be introduced to deal with these and other complications (e.g., the perceptible nonideality of any real gas). Nevertheless, with the indicated simplifying assumptions we can readily obtain calculated results quite good enough to demonstrate—by their concordance with actual experimental values—the essential soundness of our theoretical analyses.

Beginning with the translational motions, we note that the spacing of the translational energy levels is of the order of 10^{-33} erg. With the Boltzmann constant $k = 1.38 \times 10^{-16}$ erg/°K, it is clear that, even at low temperatures, the magnitude of kT is much greater than the spacing of the translational energy levels. The translational degrees of freedom will then be fully excited, as essentially classical degrees of freedom, all the way down to *extremely* low temperatures (but not in the actual limit $T \rightarrow 0$). Thus we conclude that $\frac{3}{2}R$ will properly represent the translational contribution to the molar heat capacities of all gases at all temperatures at which they can be handled *as gases*. And indeed, no gaseous heat capacity falling short of $\frac{3}{2}R$ has yet been observed.

Turning then to the rotational degrees of freedom, we confront a situation of *almost* the same simplicity. The energy spacing between the rotational levels is much greater (of the order of 10^{-16} erg) and, as noted earlier, reaches a maximum in molecular hydrogen. Hydrogen is actually *doubly* exceptional: the energy spacing between its rotational levels is conspicuously high *and* its boiling point is conspicuously low. Given this pair of properties, with gaseous hydrogen we should be able to reach a temperature quite low enough that the rotational degrees of freedom will remain completely unexcited (i.e., rotational levels above the ground state remain wholly untenanted). In these circumstances we may expect to observe a heat capacity (C_V) just equal to the $\frac{3}{2}R$ contributed by excitation of the translational degrees of freedom *only*. As the temperature is raised toward the characteristic temperature $\theta_r (\simeq 85°K$ for hydrogen) we may expect a progressive excitation of the rotational degrees of freedom, and we should observe a correspondingly progressive increase of C_V above the minimum value of $\frac{3}{2}R$. Finally, as the temperature is raised well above θ_r, we may expect that the rotational degrees of freedom will become fully activated, and we should observe a leveling-off of the total heat capacity at $\frac{5}{2}R(= \frac{3}{2}R$ translational contribution $+ \frac{2}{2}R$ rotational contribution). Now all these predictions are most satisfactorily confirmed by the empirical heat-capacity curve for hydrogen, shown in Fig. 24.† Recall however that all these predictions arise from the fact that hydrogen is *doubly* exceptional. In all *other* gases the rotational degrees of freedom should and do remain fully excited even at the lowest temperatures compatible with preservation of the gaseous state. Thus, for all diatomic gases except hydrogen, we can take $\frac{2}{2}R = R$ to represent the rotational contribution to the total molar heat capacity.

Carrying forward a total of $\frac{5}{2}R$ as the *constant* sum of translational and rotational contributions to the heat capacities of all ordinary diatomic

---

† The further rise of C_V beyond 500°K is due to excitation of the vibrational degree of freedom, of which we say more presently.

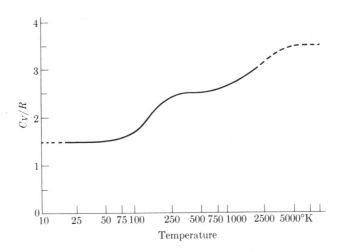

Figure 24

gases, we turn now to vibrational contributions that are more highly *variable*. The energy spacing of the vibrational levels is of the order of 10^{-13} erg—approximately 1000 times the spacing of the rotational levels, and vastly greater than the spacing of the translational states. Thus the vibrational degree of freedom may remain completely unexcited even at relatively high temperatures. In equation (26), the value of the Einstein function (in brackets) increases from *zero* at $T \to 0$ (when $\theta_v \gg T$) toward *one* at high T (when $\theta_v \ll T$). And the vibrational contribution to the molar heat capacity of a diatomic gas increases, correspondingly, from 0 at low temperature toward a limiting value of $Nk = R$ at high temperature. The limiting value is constant, but the temperature at which it is attained varies widely from substance to substance characterized by very different values of $\theta_v [= h\nu/k = (h/k)(1/2\pi)\sqrt{\kappa/\mu}]$. For a molecule involving light atoms firmly linked by a bond with high force constant, we have κ large, μ small, and θ_v very high—and the vibrational degree of freedom may remain wholly unexcited even at temperatures well above room temperature. On the other hand, for a molecule involving heavy atoms loosely linked by a bond with low force constant, we have κ small, μ large, and θ_v relatively low—and the vibrational degree of freedom may be fully excited at room temperature. In each actual case the magnitude of θ_v can be established by spectroscopic determination of the vibrational frequency (ν) characterizing the substance in question. For hydrogen θ_v is conspicuously high ($\simeq 6100°K$) and, as shown in Fig. 24, the vibrational degree of freedom for H_2 begins to be excited only well above room temperature. Even for such gases as N_2 and O_2 we see from the following table that the values of C_V at room temperature differ but little from the value $\frac{5}{2}R$ (= 4.98 cal/mole-°K) representing the translational and rotational contribution *only*. However, as we come to gases formed from progressively heavier atoms, we find a progressive growth of C_V. In iodine vapor we at last encounter a substance in which the vibrational degree of freedom is almost fully excited even at room temperature, so that the heat capacity comes close to the limiting value $\frac{7}{2}R$ (= $\frac{3}{2}R$ translational + $\frac{2}{2}R$ rotational + R vibrational) = $\frac{7}{2}(1.99)$ = 6.97 cal/mole-°K.

C_V, cal/mole-°K at 25°C (298°K)							
Gas	H_2	N_2	O_2	F_2	Cl_2	Br_2	I_2
C_V	4.90†	4.97	5.03	5.53	6.12	6.61	6.82

All this is just as it should be. Moreover, we can easily attain a gratifying unification of a great many data that express, for a great many different diatomic gases, the variation of molar heat capacity with temperature. Using spectroscopic measurements to establish the parameter θ_v characteristic of each gaseous species, we prepare for each gas a graphical display of C_V as ordinate vs the quotient T/θ_v as abscissa. Since this quotient is the *only* variable in our heat-capacity equation (26), *all* points for *all* gases should fall on the *one* curve representing the progressive excitation of a mole of harmonic oscillators. The results plotted in Fig. 25 demonstrate that this expectation is most satisfactorily confirmed by actual experimental data. We can then easily go one step further—to calculate the value of C_V for a given gas at a given temperature. Chlorine at room temperature presents an interesting problem, in that we must determine the contribution of a *partially* excited vibrational degree of freedom. Waiving consideration of the slight complication presented by the presence in this gas of isotopically distinct species—$Cl^{35}Cl^{35}$, $Cl^{35}Cl^{37}$, and $Cl^{37}Cl^{37}$—we use as the (spectroscopically determined) characteristic vibration frequency $v = 1.69 \times 10^{13}$ sec^{-1}. Then

$$\theta_v = \frac{hv}{k} = \frac{6.63 \times 10^{-27} \times 1.69 \times 10^{13}}{1.38 \times 10^{-16}} = 812°K.$$

At 298°K, therefore, $\theta_v/T = 812/298 = 2.72$. The vibrational contribution to C_V may now be determined by direct substitution in equation (26) or, more simply, by consulting a tabulation of figures for the (bracketed)

---

† Observe that in this case the rotational degree of freedom is still not *quite* fully excited even at room temperature.

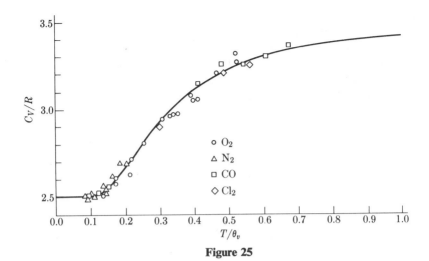

Figure 25

Einstein function. By either means, for $\theta_v/T = 2.72$ we find that

$$(C_V)_{\text{vib}} = 0.56R.$$

But then we have:

$$C_V = (C_V)_{\text{trans}} + (C_V)_{\text{rot}} + (C_V)_{\text{vib}}$$

$$= \tfrac{3}{2}R + \tfrac{2}{2}R + 0.56R = 3.06R = 6.09 \text{ cal/mole-}°K,$$

which is in excellent agreement with the experimental value of 6.12.

THE EQUILIBRIUM STATE

In an ideal gas mixture, the mutual independence of every component is guaranteed by definition. And even with a mixture of real gases (under modest pressure) we can *separately* evaluate the partition function of each component, just as if it were the only substance present. On the other hand, concerned as we now are with the *relation* of thermodynamic magnitudes for different species, we must take care consistently to base all our partition-function calculations on the same SSA reference zero of energy. Thus we scrupulously distinguish $\tilde{z}_{\text{vib}}$ based on the SSA zero from z_{vib} based on $\epsilon_0 \equiv 0$ and, in the same way, we distinguish $\tilde{Z}_{\text{tot}}$ incorporating a $\tilde{z}_{\text{vib}}$ term from Z_{tot} incorporating a z_{vib} term.

For any one component in a mixture of ideal gases, we may combine equations (10) and (17) to obtain

$$\tilde{Z}_{\text{tot}} = \frac{1}{N!}(z_{\text{trans}})^N \cdot (z_{\text{rot}})^N \cdot (\tilde{z}_{\text{vib}})^N \cdots = \frac{1}{N!}(\tilde{z})^N,$$

where by $\tilde{z}$ we shall hereafter symbolize the *total* partition function per unit, calculated on the SSA basis. Drawing then on equation (20), we find for the component's total Helmholtz free energy

$$A = -kT \ln \tilde{Z}_{tot} = -kT \ln \frac{1}{N!} (\tilde{z})^N = -kT \ln \tilde{z}^N + kT \ln N!$$

As on p. 68, application of Stirling's approximation yields

$$A = -NkT \ln \tilde{z} + kT(N \ln N - N) = -NkT(\ln \tilde{z} - \ln N) - NkT.$$

For one mole of units, we set N equal to Avogadro's number (N). Recalling that $Nk = R$, we at once obtain a suggestive expression involving the total Helmholtz free energy *per mole*

$$\bar{A} + RT = -RT \ln (\tilde{z}/N).$$

Now the Gibbs free energy (G) is related to the Helmholtz free energy by the equation $G = A + PV$. For one mole of ideal gas, described by the equation of state $P\bar{V} = RT$, we then obtain $\bar{G} = \bar{A} + RT$. Consequently

$$\bar{G} = -RT \ln (\tilde{z}/N). \tag{28}$$

By $\tilde{z}^0$ let us symbolize the value (soon to be calculated) of the partition function under some *standard* condition: e.g., at 1-atm pressure. The standard Gibbs free energy per mole ($\bar{G}^0$) will now be expressible as

$$\bar{G}^0 = -RT \ln (\tilde{z}^0/N)$$

or, for n moles,

$$G^0 = n\bar{G}^0 = -nRT \ln (\tilde{z}^0/N) = -RT \ln (\tilde{z}^0/N)^n.$$

Consider now the general type reaction:

$$aA + bB = lL + mM.$$

Using subscript letters to indicate each of the species concerned, and assuming that all species are reasonably approximated as ideal gases, we find for the overall reaction

$$\Delta G^0 \equiv l\bar{G}_L^0 + m\bar{G}_M^0 - a\bar{G}_A^0 - b\bar{G}_B^0$$

$$= -RT \ln \frac{(\tilde{z}_L^0/N)^l (\tilde{z}_M^0/N)^m}{(\tilde{z}_A^0/N)^a (\tilde{z}_B^0/N)^b}.$$

But from thermodynamics we well know that

$$\Delta G^0 = -RT \ln K_p,$$

where K_p is the equilibrium constant expressed in the same units (e.g., atmospheres) used in defining the standard state to which the $\bar{G}^0$'s refer.

Together, the last two equations enforce the conclusion that

$$K_p = \frac{(\tilde{z}_L^0/N)^l(\tilde{z}_M^0/N)^m}{(\tilde{z}_A^0/N)^a(\tilde{z}_B^0/N)^b} \tag{29}$$

From partition functions we can then at once pass to calculation of the equilibrium constant K_p for *any* gas-phase reaction.

An often useful alternative expression for K_p can be written in terms of a partition function z^0, by which we shall hereafter symbolize the *total* partition function per unit calculated relative to the ground-state energy $\epsilon_0 \equiv 0$ for the standard condition of, say, 1-atm pressure. As defined by equation (10), z^0 will differ from $\tilde{z}^0$ only in that z^0 contains a z_{vib}^0 term where $\tilde{z}^0$ contains a $\tilde{z}_{vib}^0$ term. But from equation (27) we know that $\tilde{z}_{vib}^0 = z_{vib}^0 e^{+\epsilon_d/kT}$. Hence it follows that

$$\frac{\tilde{z}^0}{N} = \frac{z^0}{N} e^{+\epsilon_d/kT}, \tag{b}$$

and substitution in equation (29) will yield

$$K_p = \frac{(z_L^0/N)^l(z_M^0/N)^m}{(z_A^0/N)^a(z_B^0/N)^b} e^{+(l\epsilon_{dL} + m\epsilon_{dM} - a\epsilon_{dA} - b\epsilon_{dB})/kT}. \tag{c}$$

The last exponent can easily be recast in a form at once more compact and more significant.

Consider the reaction as (hypothetically) it would occur with all the reactants and products in their standard states at $0°K$. All units are then in their respective ground states, with translational and rotational energies effectively equal to zero on the SSA scale. But, as we saw on p. 95, in its vibrational ground state each molecule of some species Y will possess an energy $-\epsilon_{dY}$ on the SSA scale. Let us now imagine the reactants dissociated into their·component atoms which, at $0°K$, would be in precisely the state we have used to define zero on the SSA scale of energies. The dissociation of the a molecules of A and the b molecules of B clearly demands an energy *input* $= +(a\epsilon_{dA} + b\epsilon_{dB})$. Let us then imagine the atoms, at zero on the SSA energy scale, combined to form the product molecules. Formation of l molecules of L and m molecules of M clearly involves an energy *output* $= -(l\epsilon_{dL} + m\epsilon_{dM})$. As is evident from the Hess-law cycle sketched in Fig. 26, the *sum* of the energy input and energy output must represent the *net* energy change $(\Delta\epsilon_0^0)$ when a *molecules* of A and b of B react to form l *molecules* of L and m of M. Consequently:

$$\Delta\epsilon_0^0 = +(a\epsilon_{dA} + b\epsilon_{dB}) - (l\epsilon_{dL} + m\epsilon_{dM})$$
$$= -(l\epsilon_{dL} + m\epsilon_{dM} - a\epsilon_{dA} - b\epsilon_{dB}).$$

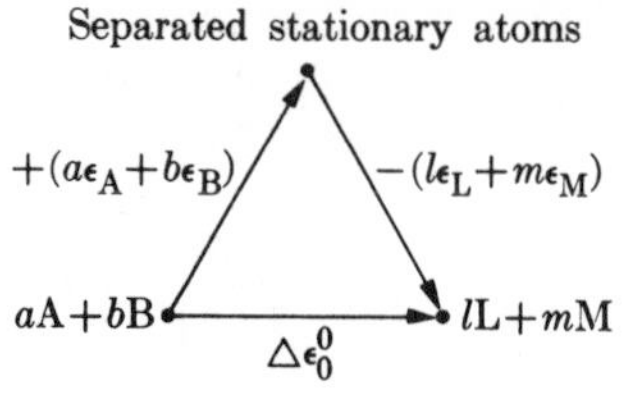

Separated stationary atoms

Figure 26

For the corresponding energy change (ΔE_0^0) in the reaction of *a moles* of A with *b* of B, forming *l moles* of L and *m* of M, we of course have $\Delta E_0^0 = N \Delta \epsilon_0^0$, so that $\Delta \epsilon_0^0 = \Delta E_0^0/N$. Dividing through the last equation above by $-kT$, and recalling that $Nk = R$, we can then write:

$$+ \frac{(l\epsilon_{dL} + m\epsilon_{dM} - a\epsilon_{dA} - b\epsilon_{dB})}{kT} = -\frac{\Delta \epsilon_0^0}{kT} = -\frac{\Delta E_0^0}{RT}.$$

But the left side of this equation is identical with the long terminal exponent in equation (c) which may, therefore, now be rewritten as:

$$K_p = \frac{(z_L^0/N)^l (z_M^0/N)^m}{(z_A^0/N)^a (z_B^0/N)^b} e^{-\Delta \epsilon_0^0/kT} = \frac{(z_L^0/N)^l (z_M^0/N)^m}{(z_A^0/N)^a (z_B^0/N)^b} e^{-\Delta E_0^0/RT}. \tag{30}$$

In any calculation of equilibrium constants from partition functions, we may choose between equations (29) and (30) whichever seems best suited to the problem in hand. We will then of course need specific expressions for either $\tilde{z}^0/N$ or z^0/N in terms of the molecular parameters characteristic of the species of units concerned. By definition

$$\frac{\tilde{z}^0}{N} = \frac{1}{N} (z_{trans}^0 \cdot z_{rot}^0 \cdot \tilde{z}_{vib}^0 \cdot z_{elec}^0 \cdots).$$

Now our earlier analyses have demonstrated that, apart from temperature, z_{rot} and $\tilde{z}_{vib}$ are functions *solely* of such species-specific parameters as θ_r, θ_v, σ, and ϵ_d. In *none* of these do we find any term representing pressure, or any variable dependent on pressure. At a given temperature, therefore, z_{rot} and $\tilde{z}_{vib}$ must be wholly independent of pressure—as is also the partition function (e.g., z_{elec}) for any other *internal* degree of freedom possessed by the ideal-gas units in question. And, given this insensitivity to changes of pressure, the last equation can at once be simplified to

$$\frac{\tilde{z}^0}{N} = \frac{1}{N} (z_{trans}^0 \cdot z_{rot} \cdot \tilde{z}_{vib} \cdot z_{elec} \cdots).$$

A quite different situation emerges when we turn to z_{trans}. For equation (21) shows that z_{trans} involves a multiplicative term (V) that is unquestionably a function of pressure. To establish the value of z_{trans}^0, we must replace this V by the volume (V^0) occupied by the system under the standard pressure (P^0) we most often take to be 1 atm. That is, equation (21) will now be written as

$$z_{trans}^0 = \left[\frac{2\pi mkT}{h^2}\right]^{3/2} V^0.$$

But for one mole of ideal-gas units $V^0 = RT/P^0 = RT/1 = RT$. Hence

$$z_{trans}^0 = \left[\frac{2\pi mkT}{h^2}\right]^{3/2} RT,$$

and substitution for z_{trans}^0 in the last equation of the preceding paragraph now yields

$$\frac{\tilde{z}^0}{N} = \left[\frac{2\pi mkT}{h^2}\right]^{3/2} \frac{RT}{N} z_{rot} \cdot \tilde{z}_{vib} \cdot z_{elec} \cdots$$

Multiplying numerator and denominator within the bracket by Avogadro's number (N), and recalling that Nm will represent the molecular weight (M) of the species in question, we rewrite the last equation as

$$\frac{\tilde{z}^0}{N} = \left[\frac{2\pi MkT}{Nh^2}\right]^{3/2} \frac{RT}{N} z_{rot} \cdot \tilde{z}_{vib} \cdot z_{elec} \cdots$$

$$= \left\{\left[\frac{2\pi k}{h^2}\right]^{3/2} \frac{R}{N^{5/2}}\right\} M^{3/2} T^{5/2} z_{rot} \cdot \tilde{z}_{vib} \cdot z_{elec} \cdots$$

If we adopt 1 atm as the *standard* pressure, and choose to express all other terms in cm-gm-sec units, the appropriate value of R is shown by **problem 18** to be 82.05 cm³-atm/mole-°K. Proceeding then to actual numerical substitutions for all the constants within the braces, we arrive at

$$\frac{\tilde{z}^0}{N} = \frac{M^{3/2} T^{5/2}}{39.0} \cdot z_{rot} \cdot \tilde{z}_{vib} \cdot z_{elec} \cdots$$

Let us now substitute in the last equation the expressions earlier obtained for z_{rot}, $\tilde{z}_{vib}$, and z_{elec}. We so obtain

$$\frac{\tilde{z}^0}{N} = \frac{M^{3/2} T^{5/2}}{39.0} \cdot \frac{T}{\sigma\theta_r} \cdot \frac{e^{+\epsilon_d/kT}}{1 - e^{-\theta_v/T}} \cdot \omega_0. \tag{31}$$

Drawing then on equation (b), we easily pass to the corresponding expression

$$\frac{z^0}{N} = \frac{M^{3/2} T^{5/2}}{39.0} \cdot \frac{T}{\sigma\theta_r} \cdot \frac{1}{1 - e^{-\theta_v/T}} \cdot \omega_0. \tag{32}$$

Observe that, in both the last equations, the dots separate the translational, rotational, vibrational, and electronic components—in that order. And both equations are very convenient forms to use in actual calculations of the equilibrium constant K_p. The values of θ_r and θ_v needed for such calculations can most readily be obtained, from spectroscopic data, if we lump together all the numerical constants appearing in the equations that define θ_r and θ_v. If once again we choose to express all terms in cm-gm-sec units, actual substitution will yield

$$\theta_r \equiv \frac{h^2}{8\pi^2 I k} = \frac{40.3 \times 10^{-40}}{I},$$

$$\theta_v \equiv \frac{h\nu}{k} = \frac{\nu}{2.08 \times 10^{+10}}.$$

Let us now bring our analytical tools to bear on a specific chemical equilibrium.

The dissociation of iodine. Consider the general equation for the dissociation of a gaseous diatomic element into its constituent atoms:

$$X_2(g) \rightleftharpoons 2X(g).$$

On the strength of equation (29), we can formulate the equilibrium constant for this reaction as:

$$K_p = \frac{(\tilde{z}_1^0/N)^2}{\tilde{z}_2^0/N},$$

where the subscripts 1 and 2 signify the species X and X_2 respectively.

For the monatomic species we need not of course consider any rotational or vibrational degrees of freedom; in addition to the translational degrees of freedom, we need allow only for the possibility of a degenerate electronic ground state. Thus, on the basis of equation (31) we can write:

$$\frac{\tilde{z}_1^0}{N} = \frac{M_1^{3/2} T^{5/2}}{39.0} \cdot (\omega_0)_1.$$

For the diatomic species X_2 we write, again on the basis of equation (31):

$$\frac{\tilde{z}_2^0}{N} = \frac{M_2^{3/2} T^{5/2}}{39.0} \cdot \frac{T}{\sigma \theta_r} \cdot \frac{e^{+\epsilon_d/kT}}{1 - e^{-\theta_v/T}} \cdot (\omega_0)_2.$$

Substituting these values in the last equation of the preceding paragraph, we find

$$K_p = \frac{\left[\dfrac{M_1^{3/2} T^{5/2}}{39.0} \cdot (\omega_0)_1\right]^2}{\dfrac{M_2^{3/2} T^{5/2}}{39.0} \cdot \dfrac{T}{\sigma \theta_r} \cdot \dfrac{e^{+\epsilon_d/kT}}{1 - e^{-\theta_v/T}} \cdot (\omega_0)_2},$$

which, on rearrangement, yields:

$$K_p = \frac{M_1^3}{M_2^{3/2}} \cdot \frac{T^{3/2}}{39.0} \cdot \frac{(\omega_0)_1^2}{(\omega_0)_2} \cdot \sigma\theta_r \cdot (1 - e^{-\theta_v/T}) \cdot e^{-\epsilon_d/kT}. \tag{d}$$

Equation (d) is a general equation for this kind of dissociation reaction, and we apply it now to the specific case of iodine:

$$I_2(g) \rightleftharpoons 2I(g)$$

at the particular temperature of 1000°C (1273°K). Spectroscopic measurements yield the following numerical data:

$(\omega_0)_1 = 4$ (i.e., the electronic ground state of the atom is fourfold degenerate),

$(\omega_0)_2 = 1$ (i.e., the electronic ground state of the molecule is non-degenerate),

$\sigma = 2$ (i.e., the symmetry factor for the molecule is 2),

$I = 750 \times 10^{-40}$ gm · cm², so that

$$\theta_r = \frac{40.3 \times 10^{-40}}{750 \times 10^{-40}} = 0.0537;$$

$v = 6.41 \times 10^{12}$ sec^{-1}, so that

$$\theta_v = \frac{6.41 \times 10^{12}}{2.08 \times 10^{10}} = 308, \text{ and}$$

$$(1 - e^{-\theta_v/T}) = 1 - e^{-308/1273} = 1 - e^{-0.242}$$

$$= 1 - 0.785 = 0.215;$$

$\epsilon_d = 2.466 \times 10^{-12}$ ergs, so that

$$e^{-\epsilon_d/kT} = e^{-2.466 \times 10^{-12}/1.38 \times 10^{-16} \times 1273}$$

$$= e^{-14.03} = \frac{1}{1.24 \times 10^6}.$$

Knowing that the atomic weight of iodine is 127 (and the molecular weight of I_2 is 254), we proceed systematically to substitute in equation (d) the values given and calculated above:

$$K_p = \frac{(127)^3}{(254)^{3/2}} \cdot \frac{(1273)^{3/2}}{39.0} \cdot \frac{(4)^2}{1} \cdot (2)(0.0537)(0.215) \frac{1}{1.24 \times 10^6}$$

$$= 0.175 \text{ atm}.$$

The experimental value for this equilibrium constant at 1000°C is 0.165 atm.   The agreement of calculated and experimental values is really excellent—and all the more delightful in that our calculation makes no allowance for nonideality of the gaseous species, anharmonicity of the vibrations (which is apt to be marked at these high temperatures), partial coupling of the rotations and vibrations, etc., etc.

Having successfully found the equilibrium constant for one particular temperature, we ask now what light statistical mechanics throws on the variation of the equilibrium constant with temperature.   Returning to equation (d), we take the logarithms of both sides, to obtain:

$$\ln K_p = \ln \left[\frac{M_1^3}{M_2^{3/2}} \frac{\sigma \theta_r}{39.0} \frac{(\omega_0)_1^2}{(\omega_0)_2} \right] + \ln T^{3/2} + \ln (1 - e^{-\theta_v/T}) + \ln e^{-\epsilon_d/kT}.$$

The first logarithmic term on the right is a collection of parameters wholly independent of temperature; this term must vanish when we differentiate with respect to temperature—as we do now:

$$\frac{d \ln K_p}{dT} = \frac{d}{dT} \left[\frac{3}{2} \ln T + \ln (1 - e^{-\theta_v/T}) - \frac{\epsilon_d}{kT} \right]$$

$$= \frac{3}{2} \frac{1}{T} - \frac{\theta_v}{T^2} \frac{e^{-\theta_v/T}}{1 - e^{-\theta_v/T}} + \frac{\epsilon_d}{kT^2}.$$

This seems a disappointingly unilluminating expression, but it can easily be recast in a more suggestive form.   Recall that ϵ_d is the energy required to dissociate one molecule into two atoms at 0°K—all species being in their standard states.   Hence $N\epsilon_d = \Delta E_0^0$, where ΔE_0^0 is the (positive) molar change in internal energy when the reaction takes place at 0°K. Substitution and rearrangement then yields:

$$\frac{d \ln K_p}{dT} = \frac{3}{2} \frac{1}{T} - \frac{\theta_v}{T^2} \frac{1}{e^{+\theta_v/T} - 1} + \frac{N\epsilon_d}{NkT^2}$$

$$= \frac{\frac{3}{2}}{T} - \frac{1}{T^2} \frac{\theta_v}{e^{+\theta_v/T} - 1} + \frac{\Delta E_0^0}{RT^2}$$

$$= \frac{\frac{3}{2}RT - R \cdot \theta_v/(e^{+\theta_v/T} - 1) + \Delta E_0^0}{RT^2}. \qquad (e)$$

The denominator in the last expression looks *very* familiar, but what are we to make of the numerator?

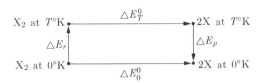

Figure 27

We begin by seeking the relation between ΔE_0^0 and ΔE_T^0, where ΔE_T^0 symbolizes the molar change in internal energy consequent to the reaction at any temperature T. Consider the Kirchhoff-law cycle sketched in Fig. 27. Since the changes of internal energy along the two routes must be the same,

$$\Delta E_r + \Delta E_T^0 + \Delta E_p = \Delta E_0^0 \qquad \text{or} \qquad \Delta E_T^0 = \Delta E_0^0 - \Delta E_r - \Delta E_p.$$

Now ΔE_r is simply the energy acquired by one mole of X_2 when it is heated (at constant volume) from $0°$ to $T°K$. At $0°K$ the translational and rotational energies are zero, and all the molecules are in the vibrational ground state. At $T°K$, on the other hand, the molecules possess translational energy $\frac{3}{2}RT$ (see p. 68) and rotational energy $\frac{2}{2}RT$ (see p. 85), as well as a vibrational energy that (as noted on p. 90) exceeds the vibrational ground-state energy by the margin of $R\theta_v/(e^{+\theta_v/T} - 1)$. Thus the *total* energy acquired by the mole of X_2 when it is heated from $0°$ to $T°K$ will be:

$$\Delta E_r = \frac{5}{2}RT + \frac{R\theta_v}{e^{+\theta_v/T} - 1}.$$

Turning then to ΔE_p, we see that it represents the energy loss suffered by the two moles of X atoms when they are cooled (at constant volume) from $T°$ to $0°K$. At temperature T they possess a translational energy of $2 \times \frac{3}{2}RT$; at $0°K$ their translational energy is effectively zero. Therefore

$$\Delta E_p = -3RT.$$

Substituting these values for ΔE_r and ΔE_p in the last equation of the preceding paragraph, we obtain:

$$\Delta E_T^0 = \Delta E_0^0 - \left[\frac{5}{2}RT + \frac{R\theta_v}{e^{+\theta_v/T} - 1}\right] - [-3RT]$$

$$= \Delta E_0^0 + \frac{1}{2}RT - \frac{R\theta_v}{e^{+\theta_v/T} - 1}.$$

How is ΔE_T^0 related to ΔH_T^0, the standard enthalpy change in the reaction at the temperature T? By definition

$$H \equiv E + PV,$$

so that

$$\Delta H_T^0 = \Delta E_T^0 + \Delta(PV)_T.$$

For an isothermal reaction involving (nominally) ideal gases, we can then write:

$$\Delta H_T^0 = \Delta E_T^0 + (\Delta n)RT,$$

where Δn is the net change in mole number. In the dissociation of X_2, one mole of reactant yields two moles of product, so that $\Delta n = 1$. This means that

$$\Delta H_T^0 = \Delta E_T^0 + RT.$$

Substituting now the expression for ΔE_T^0 obtained at the end of the preceding paragraph, we find:

$$\Delta H_T^0 = \frac{3}{2} RT - \frac{R\theta_v}{e^{+\theta_v/T} - 1} + \Delta E_0^0. \tag{f}$$

Comparing the right side of this equation with the numerator on the right side of equation (e) we see that they are *identical*. Substitution of (f) in (e) then yields:

$$\frac{d \ln K_p}{dT} = \frac{\Delta H_T^0}{RT^2}. \tag{g}$$

And so we have derived, for the special case here in question, a general theorem well known in thermodynamics as van't Hoff's law.

With equation (f) in hand, we may go on to determine, from our spectroscopic data, the actual value of ΔH_T^0 in the special case of the dissociation of I_2 at 1000°C. Recall that we earlier established the magnitude of θ_v as 308. And, with $\epsilon_d = 2.466 \times 10^{-12}$ ergs, we can at once write that $\Delta E_0^0 = 6.02 \times 10^{23} \times 2.466 \times 10^{-12}/4.18 \times 10^7 = 35,480$ cal (where the divisor 4.18×10^7 converts ergs to calories). Substituting these values in equation (f), taking $R = 1.99$ cal, and T as 1273°K, we obtain:

$$\Delta H_T^0 = \frac{3}{2}(1.99)(1273) - \frac{(1.99)(308)}{e^{+308/1273} - 1} + 35,480$$

$$= \qquad 3800 \qquad - \qquad 2230 \qquad + 35,480$$

$$\simeq 37,000 \text{ cal}.$$

Observe that the (first two) temperature-dependent terms on the right are individually small compared to ΔE_0^0 and that, moreover, those terms are so opposed in sign that they largely cancel each other. As in many other instances, we thus find here that ΔH_T^0 is very little different from ΔE_0^0—which means a ΔH_T^0 nearly invariant with temperature. The origin of this invariance is easily found in the solution to **problem 29(a)**.

Once assured of the relative constancy of ΔH^0, we can use the integrated form of equation (g) to extend to other temperatures our statistical calculation of the equilibrium constant for the reaction $I_2(g) \rightleftharpoons 2I(g)$. Treating ΔH^0 as constant, by integration of (g) we obtain:

$$\ln \frac{K_2}{K_1} = -\frac{\Delta H^0}{R}\left(\frac{1}{T_2} - \frac{1}{T_1}\right).$$

For $T_1 = 1273°K$, our earlier calculation has yielded the corresponding value $K_1 = 0.175$ atm. Inserting also our calculated value for ΔH^0 at 1273°K, and shifting over to denary logarithms, we find:

$$\log \frac{K_2}{0.175} = -\frac{37,000}{(2.30)(1.99)}\left(\frac{1}{T_2} - \frac{1}{1273}\right) = -\frac{8070}{T_2} + 6.343.$$

Dropping the now unnecessary subscript 2, we take T to be 1200°C (1473°K). Then

$$\log \frac{K}{0.175} = -\frac{8070}{1473} + 6.343 = 0.864,$$

$$\frac{K}{0.175} = 7.31, \quad K = 1.28 \text{ atm.}$$

The experimental value of K at this temperature is 1.23 atm. As one more case, let us take T to be 800°C (1073°K). Then

$$\log \frac{K}{0.175} = -\frac{8070}{1073} + 6.343 = -1.179 = +0.821 - 2,$$

$$\frac{K}{0.175} = 6.62 \times 10^{-2}, \quad K = 0.0116 \text{ atm.}$$

The experimental value for K at this temperature is 0.0114 atm.

These successes may seem entirely routine, and hardly worth achieving. After all, we know from thermodynamics that van't Hoff's law must "work." However, the above calculations embody a major element of novelty: the value of ΔH^0 we have used is *not* a calorimetric value but, rather, one we have ourselves *calculated* from spectroscopic data. To this extent the successful outcome of our analysis was *not* a foregone conclusion. Thus the striking success actually achieved represents an impressive confirmation of the validity of our statistical calculation of ΔH^0—and of our entire deployment of partition functions in the statistical calculation of equilibrium constants.

Determinants of equilibrium. For *any* reaction, on p. 43 we drew from thermodynamics the following expression for the equilibrium constant:

$$\ln K_p = -\frac{\Delta H_T^0}{RT} + \frac{\Delta S_T^0}{R}.$$

For any *gas-phase* reaction, the corresponding expression drawn from our statistical analysis is equation (30), now rewritten in logarithmic form as

$$\ln K_p = -\frac{\Delta E_0^0}{RT} + \frac{1}{R}\left[R \ln \frac{\prod (z_p^0/N)^p}{\prod (z_r^0/N)^r} \right]. \tag{33}$$

The $\prod$-symbols indicate that in the numerator we are to multiply together the series of $(z_p^0/N)^p$ terms written for all the products (p) and, in the denominator, the series of $(z_r^0/N)^r$ terms written for all the reactants (r). Whenever ΔE_0^0 is the overwhelming contributor to the magnitude of ΔH_T^0—as is most often true—the value of ΔS_T^0 will be well represented by the term in brackets in equation (33). We can then easily carry out for the *general* case the same kind of analysis earlier made of the simple reaction $A(g) \rightleftharpoons B(g)$.

As a chemical reaction progresses, energy that was originally spread only over the quantum states of the reactants becomes still further spread over the quantum states of the products. At all reasonably low temperatures the first term in equation (33), $-\Delta E_0^0/RT$, will be controlling. Thus at low temperature equilibrium must favor whichever side of the reaction is reached with evolution of energy, and the more negative the value of ΔE_0^0, the larger will be the positive magnitude of K_p for the exothermic reaction. For of course the exothermic reaction will yield products that have those energetically lower-lying quantum states which, in conformity with the Boltzmann distribution law, will be by far the most densely populated at low temperature. At high temperature, on the other hand, the first term (with T in the denominator) must ultimately become negligible. The second term, the partition-function expression in equation (33), now assumes control. Thus dominated by the standard entropy change ΔS_T^0, equilibrium then represents the condition in which energy is spread as thinly as possible over the greatest possible number of quantum states.

Whenever the quantum states accessible to the products far outnumber the quantum states accessible to the reactants, product species will far exceed reactant species in the high-temperature equilibrium mixture. For any given species of unit, the partition function offers a quantitative measure of the number of states—each duly weighted in terms of its accessibility— and so expresses the "spread" provided by that species of unit at a given temperature. By representing the extent to which the "spread" available in the reaction products differs from the "spread" available in the reactants, the partition-function ratio in brackets in equation (33) expresses the extent to which the entropy of the products differs from the entropy of the reactants. In terms of the ratio in brackets, we can then grasp what determines the sign and magnitude of ΔS_T^0 for any given reaction of ideal gases.

Whenever $\prod (z_p^0/N)^p \gg \prod (z_r^0/N)^r$, ΔS_T^0 will be large and positive. What will produce the disparity in number of accessible quantum states that is reflected in this relation of the partition functions? The narrower the spacing between successive energy levels, and the higher their degeneracy, the greater the number of accessible quantum states. In order of increasing importance, factors thus enlarging the number of accessible states include (1) a more highly degenerate electronic ground state; (2) somewhat looser bonds, with smaller force constants κ, since

$$z_{vib} = \frac{1}{1 - e^{-\theta_v/T}}, \qquad \text{where} \qquad \theta_v = \frac{h}{k} \nu = \frac{h}{k} \frac{1}{2\pi} \sqrt{\frac{\kappa}{\mu}};$$

or (3) a less symmetric or less compact molecular structure, with a lower symmetry factor σ or a higher moment of inertia I, since

$$z_{rot} = \frac{T}{\sigma \theta_r}, \qquad \text{where} \qquad \theta_r = \frac{h^2}{8\pi^2 Ik}.$$

A comparison of spectroscopically determined molecular parameters thus yields, beyond power of prediction, power to *understand* what determines ΔS_T^0 for a given reaction—and, thence, the balance of products to reactants in the high-temperature equilibrium mixture.

In addition to the three factors just noted, there remains a fourth of much greater importance. When a reaction produces change in the number of moles of gas present, the consequent disparity of translational contributions becomes the *primary* determinant of ΔS_T^0. The energy spacing of the translational quantum states is so much narrower than the spacing of rotational and vibrational states that z_{trans} is always much larger than z_{rot} and z_{vib} under comparable conditions. For diatomic gases around room temperature, order-of-magnitude values per *one* degree of freedom are

$$z_{trans} \simeq 10^{10}, \qquad z_{rot} \simeq 10^1, \qquad z_{vib} \simeq 10^0.$$

Consider then any reaction the stoichiometry of which entails that the number of moles of gaseous products exceeds the number of moles of gaseous reactants. The surplus units of gaseous product—each with its full complement of translational quantum states—ensure that the partition-function expression for the products will exceed the partition-function expression for the reactants. Such a reaction *must* proceed with increase of entropy, and must become favored at sufficiently high temperatures. The equilibrium $I_2(g) \rightleftharpoons 2I(g)$ discussed earlier offers a typical example of this effect.

We can easily extend this line of argument to the equilibrium of a monatomic gas with a solid crystal of the same substance, in a sealed vessel of constant volume. In the solid (phase A) the atoms have only vibrational degrees of freedom, with relatively wide energy spacing between successive quantum levels. In the gas (phase B) the atoms occupy translational quantum states with comparatively narrow energy spacings. The B-phase states are thus vastly more numerous than the A-phase states. Some energy $\Delta\epsilon_0^0$ is required to transfer an atom from the vibrational ground state of the solid to the translational ground state of the vapor. At low temperatures, when $kT \ll \Delta\epsilon_0^0$, most atoms will occupy phase-A states, i.e., the solid will have a low vapor pressure. As the temperature is raised, the phase-B states will become increasingly populated, i.e., the vapor pressure will rise. At high temperatures, when $kT \gg \Delta\epsilon_0^0$, both A- and B-states will be comparatively evenly populated and, since the number of B-states vastly exceeds the number of A-states, we deduce that the crystal will be wholly vaporized. *That* crystals do vaporize at high temperature is surely no surprise. Since vaporization proceeds with increase of entropy, thermodynamics teaches us to see even *how* it is that evaporation is favored at high temperatures. But it is only statistical mechanics which, passing beyond the "that" and the "how," teaches us to see in detail *why* this is the case.

Retrospect. Here endeth our survey of statistical mechanics. We have touched upon only a very few elementary applications of what is, in fact, a science of the utmost generality. A completely rigorous construction of the foundations of statistical mechanics presents deep problems of an essentially philosophical nature—problems still not fully resolved after almost a century of work by a succession of profound scholars, beginning with L. Boltzmann. But the applicability of statistical mechanics is limited only by our present incapacity to calculate partition functions for all of the many types of chemical systems we would like to subject to this mode of analysis. Yet unlike thermodynamics, which is today a subject fully worked out in principle, statistical mechanics is still a science very actively in growth. Some of its recent successes, e.g., in the analysis of the behavior of solutions of high polymers, offer ample basis for the expectation of still greater successes to come. Even as it stands, statistical mechanics already opens to analysis vast new provinces (e.g., chemical reaction kinetics) entirely foreign to the realm of classical thermodynamics. Moreover, within that realm, the powerful methods of calculation offered us by classical thermodynamics are invested with new potency as we come to learn how thermodynamic functions can be evaluated from spectroscopic data quite independent of the purely calorimetric data that are, perforce, the primary reliance of classical thermodynamics. Finally, statistical mechanics offers us the immense intellectual satisfaction of rendering transparent what thermodynamics leaves opaque: that is, statistical mechanics casts a brilliant light of *interpretation* throughout most of the realm in which classical thermodynamics can afford us little more than phenomenological *description*.

Problems

$$\text{Planck's constant, } h = 6.626 \times 10^{-27} \text{ erg-sec}$$

$$\text{Boltzmann constant, } k = 1.381 \times 10^{-16} \text{ erg/}°\text{K}$$

$$\text{Avogadro's number, } N = 6.023 \times 10^{23} \text{ mole}^{-1}$$

$$\text{Gas-law constant, } R = 8.314 \times 10^{7} \text{ ergs/mole-}°\text{K}$$

$$= 1.987 \text{ cal/mole-}°\text{K}$$

1. The game of checkers is played on an array of 32 squares distinguished by their geometric placement. Consider a game that has reached a stage at which Black has 4 men and 4 kings while Red also has 4 men and 4 kings. Noting that the different kinds of counters (e.g., red man and red king, or red king and black king) are readily distinguishable from each other, but that counters of the same sort (e.g., the 4 red men) are not thus distinguishable, determine the number of recognizably different positions possible at this stage of the game.

2. a) Show that, in five-card stud poker, the number of possible five-card hands is

$$\frac{52 \cdot 51 \cdot 50 \cdot 49 \cdot 48}{5 \cdot 4 \cdot 3 \cdot 2 \cdot 1} = \frac{52!}{5! \, 47!} \simeq 2.6 \times 10^{6}.$$

b) Consider the probability that, in a game of five-card stud, you will be dealt a flush (i.e., five cards of the same suit). By showing that the number of ways in which you may be dealt a flush is

$$\frac{52 \cdot 12 \cdot 11 \cdot 10 \cdot 9}{5 \cdot 4 \cdot 3 \cdot 2 \cdot 1} = 4 \times \frac{13!}{5! \, 8!} \simeq 5 \times 10^{3},$$

demonstrate that your chance of being dealt a flush is of the order of $\frac{1}{500}$.

3. a) In draw poker there is about a 49% chance that a randomly dealt five-card hand will contain at least two cards of the same denomination. This conclusion is perhaps most readily demonstrated by showing that the probability that the hand does *not* contain two cards of the same denomination is given by the expression

$$\frac{52}{52} \cdot \frac{48}{51} \cdot \frac{44}{50} \cdot \frac{40}{49} \cdot \frac{36}{48}.$$

Justify this expression, and complete the indicated demonstration.

b) In a class of 50 randomly selected students, what is the probability that two will have the same birthday? This surprisingly large probability is easily obtained by adapting the development in part (a) to calculate the probability that two students will *not* have the same birthday.

4. [After Sussman, *J. Chem. Educ.* **40**, 49 (1963).] a) In the neck of the sealed flask shown below, 5 white balls rest on top of 5 black balls. Suppose that the balls are tipped into the body of the flask, and that the flask is well shaken and then returned to the inverted position shown here. Demonstrate that the probability that the indicated arrangement will again turn up is

$$\frac{5}{10} \cdot \frac{4}{9} \cdot \frac{3}{8} \cdot \frac{2}{7} \cdot \frac{1}{6} = \frac{5!\,5!}{10!} = \frac{1}{252}.$$

b) What is the physical significance of the function $10!/5!\,5!$?

c) Determine the number of distinguishable arrangements that would be possible if all ten balls were differently colored.

5. a) The *total* number of miscrostates (Ω) associated with *all* configurations possible when N distinguishable harmonic oscillators share Q energy quanta is:

$$\Omega = \sum W_i = \frac{(N + Q - 1)!}{(N - 1)!\,Q!}.$$

To derive this formula, we symbolize each of the oscillators by an *o*, each of the quanta by a *u*. At the left end of a line we set down one *o*, and then consider in how many ways we can follow the one *o*, on the same line, with the remaining $(N - 1)$ *o*'s and the Q *u*'s. Conceiving each *o* as "possessing" all *u*'s appearing to its right prior to the appearance of the next *o*, we see that the total number of permutations possible will also represent the total number of possible microstates (Ω). Justify this statement, and derive the above formula.

b) Use the formula to confirm the statement made in the text that 2002 distinct microstates are possible when 10 oscillators share 5 energy quanta.

c) Use the formula, and Stirling's approximation, to confirm that 1000 oscillators sharing 1000 energy quanta can assume a total in the order of 10^{600} microstates.

6. a) Justify the claim that the formula $W = N!/(H!)(T!)$ yields the number of ways (W) in which some particular numbers of heads (H) and tails (T) can be produced in a series of N tosses of a well-balanced coin.

b) For a series of 1000 tosses of a coin, we define $A = W_{H-T}/W_{500-500}$, where $W_{500-500}$ is the number of microstates associated with the predominant configuration, and W_{H-T} is the number of microstates associated with the configuration characterized by some particular numbers of heads (H) and tails (T). Show that $\log A = 1000 \log 500 - H \log H - T \log T$.

c) Using the formula noted in (b), show that a 400–600 configuration possesses only about 10^{-9} as many microstates as the predominant 500–500 configuration.

d) Amending the formula in (b) as required for a series of 100 tosses, show that the 40–60 configuration has (as indicated in Fig. 7) approximately $\frac{2}{15}$ as many microstates as the predominant 50–50 configuration.

e) Explain the drastic difference in the A-ratios for the 40-60 and 400-600 configurations.

f) With numbers as small as those above, the applicability of Stirling's approximation is certainly questionable. To appraise the extent of the error so introduced in the A-ratio calculated in part (d), repeat that calculation—using in place of Stirling's approximation the following accurate figures:

$$40! = 8.16 \times 10^{47}; \qquad 50! = 3.04 \times 10^{64}; \qquad 60! = 8.32 \times 10^{81}.$$

7. a) Let W_{max} symbolize the number of ways in which $N/2$ heads and $N/2$ tails may show up as the outcome of a series of N tosses of a well-balanced coin. Let W symbolize the number of ways in which these tosses may yield some other figures for heads (H) and tails (T). For N, H, and T all large numbers, show that

$$\log \frac{W}{W_{max}} = -H \log \frac{H}{N/2} - T \log \frac{T}{N/2}.$$

b) Relative to the predominant configuration in which heads equal tails, the divergence of any other configuration may be measured in terms of the index

$$\alpha \equiv \frac{H - T}{N},$$

where α ranges from -1, through 0 for the predominant configuration, up to $+1$. Show that H and T are expressible in terms of this index as

$$H = \frac{N}{2}(1 + \alpha) \qquad \text{and} \qquad T = \frac{N}{2}(1 - \alpha).$$

c) Pursuing the general line of the argument on pp. 20–21, substitute from (b) in (a) to show that—for configurations that diverge only slightly from the predominant configuration—W is related to $W_{\max}$ by a simple Gaussian function

$$\frac{W}{W_{\max}} = e^{-N\alpha^2}.$$

Indicate the bearing of this expression on the shape of the curves in Fig. 7.

8. As a description of the predominant configuration of an assembly of harmonic oscillators, we obtained the simple geometric series

$$\eta_1/\eta_0 = \eta_2/\eta_1 = \cdots \eta_{n+1}/\eta_n \cdots = \text{constant.}$$

We never considered explicitly whether the constant is a number greater or less than 1; i.e., whether the populations form an *ascending* series in which $\eta_{n+1} > \eta_n$, or a *descending* series in which $\eta_{n+1} < \eta_n$. The appearance of the predominant configuration (VI) in Fig. 5 strongly suggests a descending series, but we should check this point in another case.

a) The accompanying figure displays two of many configurations that may be assumed by an assembly of 15 oscillators sharing 20 quanta. Both these configurations conform to the geometric-series requirement as closely as possible in a small assembly like this, but configuration I represents an ascending series while configuration II represents a descending series. As evidence that the predominant configuration is a descending series, show that $W_{II}/W_I = 168/1$.

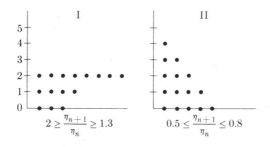

b) Is it possible that the predominant configuration is neither an ascending nor a descending series but, rather, a series of equal population numbers? To check on this possibility, consider a new configuration produced from II by moving one oscillator from level 4 to level 3 and one oscillator from level 1 to level 2. Does this approach to a more equal set of population numbers actually increase or decrease W, and by what factor?

c) We have examined only a few special cases, but we can easily see why a descending series will *in general* yield a greater value for W than does an otherwise similar ascending series. With the total energy a constant, observe that placement of many oscillators in higher levels (as in an ascending series) entails that *all* oscillators will be confined to a relatively small group of densely populated low-energy levels. This in turn means that $\prod \eta_n!$ will be comparatively large, and W will be correspondingly reduced. With these pointers, construct the general argument for the conclusion that, regardless of the species of units concerned, the predominant configuration represents a set of *decreasing* population numbers.

9. To an equilibrium assembly in its predominant configuration, let there be added some small quantity of energy (dE). As a result of this addition, let us suppose that the population of the lowest level changes from η_0 to $(\eta_0 + x_0)$, the population of the next level from η_1 to $(\eta_1 + x_1)$, and, in general, the population of the nth level from η_n to $(\eta_n + x_n)$. Some of the x's will of course be negative numbers.

a) Symbolizing by W_i and W_f respectively the numbers of microstates before and after the energy transfer, by the methods used in obtaining equation (3) show that

$$W_f/W_i = \prod \eta_n! / \prod (\eta_n + x_n)! \simeq 1/\prod (\eta_n)^{x_n}.$$

b) Drawing now on a suitable form of equation (3), show that

$$\ln W_f - \ln W_i = \sum \ln (\eta_0 e^{-\beta \epsilon_n})^{-x_n} = \sum \beta x_n \epsilon_n - \sum x_n \ln \eta_0.$$

c) Comparing this equation with the corresponding expression on p. 27, indicate the circumstances in which the last equation will reduce to

$$d \ln W = \beta \, dE.$$

Have the assumptions made in this derivation of equation (5) differed at all from the assumptions noted for the derivation in the text?

d) For a finite fractional shift (α_n) in the populations of the various quantum levels, on p. 21 we obtained the equation

$$\ln (W_{max}/W) = \sum \alpha_n \eta_n \ln \eta_0 - \sum \beta \epsilon_n \alpha_n \eta_n + \sum \eta_n \alpha_n^2.$$

Observe that $\alpha_n \eta_n = x_n$ as defined in this problem, and note also how W_{max} and W relate to W_i and W_f respectively. By a suitable combination of the first and third terms on the right-hand side in the last equation, show that in the limit $\alpha_n \to 0$ this equation reduces to the equation written under (b) above.

10. Consider an assembly of N harmonic oscillators sharing Q of those energy quanta each of which suffices to promote one oscillator to the next higher quantum state. Equation (a) on p. 16 prompts us to define a new parameter $\gamma \equiv \eta_{n+1}/\eta_n$, and for the given assembly γ will then be a *constant less than* 1. Regarding N and Q as characteristics of the assembly that are in principle determinable, we now seek to establish the actual equilibrium distribution in terms of N and Q.

a) If by η_0 we symbolize the population of the ground state for which we write $\epsilon_0 \equiv 0$, show that

$$N = \sum \eta_n = \eta_0(1 + \gamma + \gamma^2 + \cdots) = \eta_0/(1 - \gamma)$$

and, further, that

$$Q = \sum n \cdot \eta_n = \eta_0 \gamma(1 + 2\gamma + 3\gamma^2 + \cdots) = \eta_0 \gamma/(1 - \gamma)^2.$$

b) From the relations obtained in (a), show that

$$N + Q = \eta_0/(1 - \gamma)^2 \quad \text{and} \quad \gamma = Q/(N + Q) \quad \text{and} \quad \eta_0 = N^2/(N + Q).$$

c) Consider a "hot" body consisting of 500 oscillators that share 750 quanta, and a "cold" body consisting of 500 units that share 250 quanta. Joining with the definition $\gamma \equiv \eta_{n+1}/\eta_n$ the expressions for η_0 and γ given under (b), show that the equilibrium distributions in the two *separate* bodies are as shown in rows 1 and 2 of the accompanying table.

		η_0	η_1	η_2	η_3	η_4	η_5	η_6	η_7	η_8	η_9	η_{10}	η_{11}	η_{12}
1	Cold body	333	112	37	12	4	1	1	—	—	—	—	—	—
2	Hot body	200	120	72	43	26	16	9	6	3	2	1	1	1
3	Hot–cold assembly	533	232	109	55	30	17	10	6	3	2	1	1	1
4	Equilibrium assembly	499	250	125	63	32	16	8	4	2	1	—	—	—

d) Explain why, immediately after the two bodies are joined, the configuration of the united system will be as shown in row 3. To demonstrate that this is *not* an equilibrium configuration, consider the test ratios η_0/η_3 and η_3/η_6. Show that, though the Boltzmann distribution law requires that these ratios be equal, they are *not* equal for the distribution shown in row 3 of the table.

e) Given that the united assembly consists of 1000 units sharing 1000 energy quanta, use the formulas given under (b) to derive the equilibrium distribution shown in row 4 of the table. As evidence that this *is* an equilibrium distribution, show that the tests ratios η_0/η_3 and η_3/η_6 are here equal.

f) As further evidence that row 4 shows an equilibrium distribution, while row 3 does not, consider the following displacement: withdrawing two units from level 1, transfer one of them to level 0 and one of them to level 2. By what factor would this displacement change W for the configuration shown in row 3? W for the configuration shown in row 4? How do these results strengthen the conclusion that row 3 is a nonequilibrium distribution while row 4 is an equilibrium distribution?

g) For the distribution in row 3, $W \simeq 10^{570}$; for the distribution in row 4, $W \simeq 10^{600}$. Why is it correct to say that, after thermal equilibration, the odds are at least $10^{30}:1$ against reappearance of the hot-cold configuration shown in row 3? To appreciate the enormity of these odds, imagine that, after the initial equilibration, you were to observe the microstates assumed by the united assembly. Imagine that you were able to observe 10^6 complete microstates per second, and that you continued to observe, without rest, from the origin of the universe some 10^{10} years ago right up to the present day. Calculate the number of observations you would by now have made, and show that the odds are still at least $10^7:1$ *against* your having once observed any microstate of the hot-cold configuration shown in row 3.

11. a) Assuming a transfer of energy from body Y to body X, in our study of thermal equilibration we arrived at the conclusion $\beta_X \geq \beta_Y$. Had we assumed a transfer from body X to body Y, show how we could then have arrived at the conclusion $\beta_Y \geq \beta_X$.

b) The zeroth law of thermodynamics is so called because it was fully recognized only after the first and second laws were well established. The zeroth law asserts that whenever two bodies have been found to stand in equilibrium with a third (e.g., a thermometer), then those two bodies will also stand in thermal equilibrium with each other. Show how the zeroth law follows at once from the relation of β's we have found to characterize the condition of thermal equilibrium.

12. a) If you are familiar with the technique of integration-by-parts, you should have no difficulty in deriving the simple form of Stirling's approximation used in the text. Perform this derivation by justifying the following series of relations, including the indicated approximations,

$$\ln N! = \sum_{1}^{N} \ln N \simeq \int_{1}^{N} \ln N \, dN \simeq N \ln N - N.$$

b) On p. 27 we call on Stirling's approximation to derive the relation $d \ln N! = \ln N \, dN$. This relation is also derivable without any use of Stirling's approximation— simply by calling on the fundamental theorem of the differential calculus. Consider that when N is of the order of 10^{23}, even $\Delta N = 1$ is effectively $\Delta N \to 0$. Show then how the indicated relation follows from an argument that begins with

$$\frac{d \ln N!}{dN} = \lim_{\Delta N \to 0} \left[\frac{\ln (N + \Delta N)! - \ln N!}{\Delta N} \right] \simeq \frac{\ln [(N + 1)!/N!]}{1}.$$

c) To justify the conclusion that W reaches its maximum when $\bar{\alpha} = 0$, we must demonstrate not only that $dW/d\bar{\alpha} = 0$ when $\bar{\alpha} = 0$ but also that $d^2 W/d\bar{\alpha}^2 < 0$ when $\bar{\alpha} = 0$. Starting from equation (4), perform both demonstrations, and explain why the second one is necessary.

13. a) Could a calculated partition function ever have a value less than one? Why?

b) When all energies are measured relative to $\epsilon_0 \equiv 0$, the population of the ith quantum state is given by equation (9) as

$$\eta_i = \frac{N e^{-\beta \epsilon_i}}{\sum e^{-\beta \epsilon_q}}.$$

Obviously the actual population η_i must remain unchanged wherever we choose to place our reference zero of energy. Demonstrate that if all energies are measured on a scale relative to which $\epsilon_0 \equiv x$, the above equation continues to yield exactly the same result for η_i.

14. In Bose-Einstein statistics there is no limit on the number of indistinguishable units that can occupy a given quantum state. In Fermi-Dirac statistics, on the other hand, each quantum state can contain no more than one such unit. Fermi-Dirac statistics can thus be approached by considering the distribution of identical counters on a gameboard of distinguishable squares each of which can at most bear one counter.

a) Show that in Fermi-Dirac statistics the number of ways (W_i) in which η_i^* indistinguishable units can be distributed over the ω_i distinct quantum states comprised within an ith energy level is expressible as

$$W_i = \omega_i!/\eta_i^* ! \, (\omega_i - \eta_i^*)!.$$

b) Show that the dilute-gas approximation, taking $\omega_i \gg \eta_i^*$, reduces the last equation to the same result obtained in Bose-Einstein statistics, namely:

$$W_i = (\omega_i)^{\eta_i^*}/\eta_i^*!.$$

c) Explain why in Fermi-Dirac statistics, just as in Bose-Einstein statistics, the total number of ways in which *all* the units present may be distributed, over *all* the ω-fold degenerate energy levels characteristic of a given assembly, will be given as

$$W = \prod [(\omega_y)^{\eta_y^*}/\eta_y^*!].$$

15. Under the dilute-gas approximation, the limiting form of both Fermi-Dirac and Bose-Einstein statistics is the expression for W given in the text as equation (12), and rewritten as the last equation above.

a) To derive from (12) the distribution law for nondistinguishable units, adapt the argument on pp. 17–19 as indicated on p. 52, to obtain

$$\eta_i^*/\omega_i = (\eta_0^*/\omega_0)e^{-\beta\epsilon_i}.$$

Explain why this equation is equivalent to the corresponding expression for distinguishable units, given in the text as equation (8).

b) The argument on p. 53 shows that from the equation derived in (a) one obtains an equation (13), here rewritten as

$$\eta_i^* = \frac{N}{z}\,\omega_i e^{-\beta\epsilon_i},$$

where $z \equiv \sum \omega_y e^{-\beta\epsilon_y}$. Substituting in equation (12) on the strength of equation (13), and recalling that $\sum \eta_y^* = N$ and that $\sum \eta_y^*\epsilon_y = E$, show that for indistinguishable (Fermi-Dirac or Bose-Einstein) units one finds

$$\ln W = -\ln N! + \ln z^N + \beta E. \tag{x}$$

c) From the last equation show that one finds for indistinguishable units, as in the text one finds for distinguishable units, that the entropy is given by the equation

$$S = k \ln Z + k\beta E.$$

16. As an expression for W when the dilute-gas approximation applies to assemblies of indistinguishable (Fermi-Dirac or Bose-Einstein) units, equation (x) given under problem 15(b) permits us to make two illuminating comparisons with assemblies of distinguishable (Boltzmann) units.

a) To establish the generality of an important conclusion stated on p. 36, we have yet to demonstrate that an assembly of indistinguishable units will conform to the same equation (5) earlier derived for an assembly of distinguishable units. Starting from the indicated equation (x), proceed in the manner of pp. 27–28 to show that, indeed, for an assembly of indistinguishable units also

$$d \ln W = \beta\, dE.$$

b) In the text, a rather tedious argument leads to the conclusion that the number of distinct states accessible to an assembly of indistinguishable units is only $1/N!$

.times the number of such states accessible to an otherwise identical assembly of distinguishable units. Comparing the above equation (x) with the corresponding equation obtained on p. 61, for ln W in an assembly of distinguishable units, demonstrate that for otherwise identical assemblies

$$W_{\text{indist}} = \frac{1}{N!} W_{\text{dist}}.$$

17. a) Consider the relation of the area of the rectangular blocks to the area under the curve in the figure shown below. For what values of α^2 can the summation $\sum e^{-\alpha^2 x^2}$ properly be replaced by the integral $\int e^{-\alpha^2 x^2} \, dx$, and why

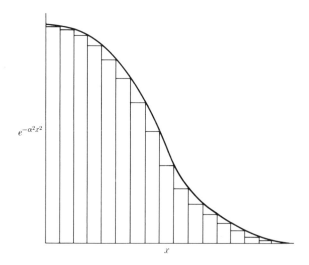

b) A rough-and-ready determination of the value (I) of the definite integral $\int_0^\infty e^{-\alpha^2 x^2} \, dx$ is easily accessible to readers unterrified by multiple integrals. Observing that x is only a dummy variable, one writes

$$I^2 = \int_0^\infty \int_0^\infty e^{-\alpha^2(x^2 + y^2)} \, dx \, dy.$$

With a shift to polar coordinates, in which the infinitesimal element of area becomes $r \, dr \, d\theta$, the foregoing integral over the positive quadrant of the xy plane becomes

$$I^2 = \int_0^{\pi/2} \int_0^\infty e^{-\alpha^2 r^2} r \, dr \, d\theta,$$

which is trivially easy to integrate. Writing out the full analysis of which only the high points have been given above, determine the function symbolized by I.

18. a) With reference to equation (21)

$$z_{\text{trans}} = \left[\frac{2\pi mkT}{h^2} \right]^{3/2} V,$$

show that when one substitutes for m, k, and h in units of gm, cm, sec, and ergs ($\equiv$ gm $\cdot$ cm^2/sec^2), one must express V in units of cubic centimeters, or milliliters.

b) Using for R the value 82.05 ml-atm/mole-°K, determine the molar volume of neon (atomic weight, 20) at 1-atm pressure and 298°K. By making appropriate numerical substitutions for all the terms in the equation for ζ appearing on p. 75, show that for a mole of neon, under the indicated conditions, ζ is of the order of 10^{30}.

c) Indicate how this result justifies the dilute-gas approximation, and explain why this value for ζ entails that the corresponding value of n shall be approximately 10^{10}, as stated on p. 74.

19. a) Starting from equation (c) on p. 71, in which -11.074 cal/mole-°K is the value of the function within the braces when V is expressed in cubic centimeters, derive the following variant of the Sackur-Tetrode equation, in which P is expressed in atmospheres:

$$\bar{S} = R \ln (M^{3/2}T^{5/2}/P) - 2.316 \text{ cal/mole-°K}.$$

b) Show that the last equation can be rewritten as

$$\bar{S} = R \ln (M^{3/2}T^{5/2}/P) + R(1 + \ln 0.115).$$

With this equation combine the crudely approximate expression obtained on p. 77,

$$\bar{S} = R \ln \frac{\zeta}{N} + R,$$

to show that

$$\frac{\zeta}{N} = 0.115 \frac{M^{3/2}T^{5/2}}{P}.$$

c) Though only *very* rough, the last expression permits some exploration of the range of applicability of the dilute-gas approximation. Appraise the applicability of this approximation to gaseous helium (atomic weight, 4) under a pressure of 1 atm at its normal boiling point of approximately 4°K.

20. a) For N molecules in a *one*-dimensional box of length L, adapt the derivation on pp. 78–79 in the text to show that, since

$$\omega = d\zeta = dn = \frac{2mL}{h} du_x,$$

the speed distribution law for this single translational degree of freedom will be

$$\frac{1}{N} \frac{dN}{du_x} = \left[\frac{2m}{\pi kT}\right]^{1/2} e^{-mu_x^2/2kT}.$$

b) Sketch and comment on the form of this function when T is large; when T is small.

c) By inspection of the last equation under part (a), derive and sketch the very closely related function that represents $(1/N)(dN/dv_x)$ where v_x is not the molecular speed but, rather, the molecular *velocity*.

21. a) By inserting the appropriate units for h, I, and k, show that $\theta_r(\equiv h^2/8\pi^2 Ik)$ has the dimensions of an absolute temperature.

b) Relative to other lines in a rotational absorption spectrum, the intensity of any given line depends primarily on the size of the population in the rotational level from which the transition originates. The population of the Jth rotational *energy* level is:

$$\eta_J^* = N\frac{2J + 1}{z_{rot}}e^{-(J)(J+1)\theta_r/T}.$$

Determine, as a function of θ_r and T, the value of J that distinguishes the most heavily populated rotational energy level.

c) In a rotation-vibration spectrum of gaseous HCl (with moment of inertia $I = 2.65 \times 10^{-40}$ gm-cm^2) the most intense absorption line originates in the level for which $J = 3$. At what temperature was the spectrum measured?

22. a) Use the expression for $(J)(J + 1)$ given on p. 86 to estimate the number of distinct rotational quantum states accessible to HCl molecules (with $I = 2.65 \times 10^{-40}$ gm-cm^2) at the normal boiling point of approximately 188°K. In these circumstances would you expect integration to offer an acceptable evaluation of the partition-function summation for the rotational degrees of freedom?

b) Let N distinguishable rotors be distributed over $(J)(J + 1)$ distinct quantum states assumed (counterfactually) to be equivalent in energy. Show that the number (W) of possible distributions is expressible as

$$W = [(J)(J + 1)]^N = \left[\frac{8\pi^2 IkT}{h^2}\right]^N.$$

c) Using the definition $S \equiv k \ln W$, show that the last equation yields, for a mole of heteronuclear rotors, the approximate result

$$\bar{S}_{rot} = R \ln IT + 175.69 \text{ cal/mole-°K}.$$

Considering the approximations involved, is this expression in reasonable agreement with the correct expression embodied in equation (24)?

23. a) If, instead of a summation, an integral is used to evaluate z_{vib}, the resulting expression will be applicable only in the limit of *very* high temperature. Why?

b) Proceeding as indicated under (a), one finds:

$$z_{vib} = \sum_0^\infty e^{-nh\nu/kT} = \int_0^\infty e^{-nh\nu/kT}\,dn = \frac{kT}{h\nu}$$

from which it follows that $\bar{E}_{vib} = RT$. Carry through the derivations that yield the indicated results for z_{vib} and $\bar{E}_{vib}$.

c) As the fraction (f) of the molecules of gaseous NO present in each vibrational state (characterized by the quantum number v), Johnston and Chapman give for 2000°K the following data:

v	0	1	2	3	4	>4
f	0.7344	0.1936	0.0521	0.0143	0.0040	0.0016

What is the energy spacing between the quantum states for which $v = 0$ and $v = 1$?

d) For 1 mole of gaseous NO at 2000°K, what is the actual value of $\bar{E}_{vib}$ implied by the data given under part (c)? Compare this with the theoretical value obtained from the equation $\bar{E}_{vib} = RT$: in this context, does 2000°K properly qualify as a "high" temperature?

24. a) The energy at the bottom of the potential-energy well of the harmonic oscillator lies $\frac{1}{2}h\nu$ below the energy (ϵ_0) of the vibrational ground state. Let z'_{vib} and E'_{vib} symbolize the partition function and vibrational energy calculated relative to $\epsilon_0 \equiv \frac{1}{2}h\nu$. Starting from the equation on p. 91 that relates z'_{vib} to z_{vib} (measured from $\epsilon_0 \equiv 0$), derive the relation of E'_{vib} to E_{vib} (measured relative to $\epsilon_0 \equiv 0$).

b) Relative to the SSA energy zero, calculate the energy (E) and Helmholtz free energy (A) per mole of linear harmonic oscillators characterized by the parameters θ_v and ϵ_d. Explain the difference between your expression for E and that given as equation (e) on p. 90.

c) From your expression for E, derive an expression for C_V; from your expressions for E and A, derive an expression for S.

d) Your formulas for C_V and S would remain unchanged even if a different energy zero had been chosen. Show how this independence is readily explicable in terms of the physical significance we associate with C_V and with S.

25. Consider an assembly of units in which the first excited electronic state is ω_1-degenerate and associated with an energy (ϵ_1) not too much greater than the energy (ϵ_0) of the ω_0-degenerate ground electronic state. Assume that all other electronic states fall at energies high enough to render them effectively inaccessible.

a) Taking as your reference zero of energy $\epsilon_0 \equiv 0$, and using ω to symbolize the *ratio* ω_1/ω_0, show that

$$z_{elec} = \omega_0(1 + \omega e^{-\epsilon_1/kT}).$$

b) For an N-unit assembly, show that

$$E_{elec} = \frac{\omega N\epsilon_1}{(e^{+\epsilon_1/kT} + \omega)}.$$

Show that $E_{elec} \rightarrow 0$ as $T \rightarrow 0$. Show further that at high temperature E_{elec} approaches a *constant*, determine the constant for units in which $\omega = 1$, and explain why your result is a reasonable one.

c) Show that

$$(C_V)_{elec} = \omega R \left(\frac{\epsilon_1}{kT}\right)^2 \frac{e^{\epsilon_1/kT}}{(e^{\epsilon_1/kT} + \omega)^2}.$$

Determine the limiting values of C_V in the extremes of very low and very high temperatures. How do you explain the high-temperature limit (i) mathematically, in terms of the difference between the above equation and equation (a) on p. 99, and (ii) physically, in terms of the distribution of these units in the high-temperature assembly?

d) Between the extremes of high and low temperatures, $(C_V)_{elec}$ reaches a maximum value. By actual differentiation of the expression for C_V (or, if you prefer, ln C_V) show that the maximum value is reached when

$$\frac{\epsilon_1}{kT} = 2.3 \left[\log \omega + \log \frac{(\epsilon_1/kT) + 2}{(\epsilon_1/kT) - 2} \right].$$

By trial-and-error substitutions in this equation, show that when $\omega = 1$ the maximum value of $(C_V)_{elec}$ is reached when $\epsilon_1/kT \simeq 2.4$.

e) Nitric oxide, NO, has $\omega_0 = \omega_1 = 2$, and $\epsilon_1 = 2.4 \times 10^{-14}$ erg. For this substance, at what temperature will $(C_V)_{elec}$ reach its maximum value, what is that value, and what is likely to be the *total* heat capacity of NO at the temperature in question?

f) Nitric oxide is *highly* exceptional in its low value of ϵ_1. Gaseous O_2 shows an $\epsilon_1 = 1.6 \times 10^{-12}$, about 70 times ϵ_1 for NO, but the value of ϵ_1 for O_2 is still considerably lower than ϵ_1 for most other diatomic gases. Calculate the temperature at which $(C_V)_{elec}$ for oxygen reaches its maximum value and, in terms of your answer, explain why we don't usually have to consider electronic excitations in our calculations of thermodynamic properties of diatomic gases.

26. This problem is designed to convey some feeling for the orders of magnitude of partition functions for a representative gaseous molecule, carbon monoxide.

a) Given 28 as the molecular weight of CO, use equation (21) to calculate the translational partition function per unit in one mole of carbon monoxide occupying 22,400 cm^3 at 273°K and 1-atm pressure.

b) Given that for CO the moment of inertia $I = 14.5 \times 10^{-40}$ gm-cm^2, calculate the rotational partition function per unit at 273°K. With due regard to the degeneracy factor, calculate also the population ratio η_1^*/η_0^* at 273°K.

c) Given that for CO the vibrational frequency $v = 6.43 \times 10^{13}$ sec^{-1}, calculate the vibrational partition function per unit at 273°K and at 1500°K. *Estimate* the population ratio η_1/η_0 at each temperature.

d) Compare your results from parts (a), (b), and (c) with the order-of-magnitude figures given on p. 120—in each case for *one* degree of freedom at a temperature of about 400°K.

27. a) By inserting the appropriate units for h, k, and v, show that the parameter $\theta_v = hv/k$ has the dimensions of an absolute temperature.

b) From Fig. 21 determine C_V for diamond at 400°K and then, using Fig. 20, determine the appropriate value of θ_v for diamond. Using this value, obtain from Fig. 20 the magnitude of C_V for diamond at 800°K. Is your result in good agreement with that indicated by Fig. 21?

c) The measured molar heat capacities of crystalline KCl at the indicated temperatures are:

T, °K	50	100	175	270
C_V, cal/mole-°K	5.05	9.31	11.03	11.61

How do you account for the fact that the limiting heat capacity here seems to approach 12 cal/mole-°K rather than the Dulong-Petit value of 6 cal/mole-°K? Will Einstein's heat-capacity theory apply to ionic compounds as well as to elements? To answer this question (i) determine θ_v for KCl from the cited value of C_V at 100°K, and (ii) use this value of θ_v to estimate C_V for KCl at the other three temperatures. Comment on the agreement of the calculated and empirical results.

28. a) In gaseous iodine the moment of inertia $I = 750 \times 10^{-40}$ gm-cm^2 and the characteristic vibrational frequency $v = 6.41 \times 10^{12}$ sec^{-1}. Calculate C_V and the entropy of 1 mole of gaseous I_2 at 298°K and 1-atm pressure—obtaining for the entropy a result approximating 62.1 cal/mole-°K.

b) Calorimetric measurements extended back toward 0°K yield for one mole of solid I_2 at 298°K the result $S_{298}^0 = 27.9$ cal/mole-°K. For the sublimation of 1 mole of I_2 at 298°K, calculate ΔS_{298}^0 as $(S_{298}^0)_{vap} - (S_{298}^0)_{sol}$.

c) At 298°K the equilibrium pressure of iodine vapor over solid iodine is 0.305 mm Hg. Use this figure (reexpressed in atmospheres) to determine ΔG_{298}^0 for the sublimation of one mole of I_2 at 298°K.

d) From the results obtained in parts (b) and (c) calculate the standard enthalpy of sublimation (ΔH_{298}^0) for one mole of I_2 at 298°K. The accepted value for this heat of sublimation is 14.88 kcal/mole.

29. a) For the reaction $I_2(g) \rightleftharpoons 2I(g)$, ΔH^0 is nearly invariant with temperature. Such invariance implies that ΔC_P (in Kirchhoff's law) must be relatively minute. On the basis of the discussion and data given on pp. 114–118, calculate $(C_P)_{I_2}$, $2(C_P)_I$, and ΔC_P for the reaction at approximately 1000°C.

b) Assuming negligibility or mutual cancellation of all contributions due to vibrational degrees of freedom, show that $\Delta H_T^0 = \Delta E_0^0$ for the reaction:

$$Br_2(g) + Cl_2(g) = 2BrCl(g).$$

c) Let us assume (i) negligibility of all vibrational contributions, (ii) $\Delta E_0^0 = \Delta H_{298}^0$ which is measured as -320 cal, (iii) approximate cancellation of all translational contributions, and (iv) moments of inertia so related that $(I_{BrCl})^2/I_{Br_2}I_{Cl_2}$ approximates unity. Making these drastic simplifications, you can easily calculate K_p and, thence, ΔG_{298}^0 for the above reaction. Is your calculated value in reasonably good agreement with the empirical finding $\Delta G_{298}^0 = -1171$ cal?

30. For the isotopic-exchange reaction

$$O^{16}O^{16} + O^{18}O^{18} \rightleftharpoons 2O^{16}O^{18}$$

the gas-phase equilibrium constant $K_p = 3.98$, and other gas-phase exchange reactions of this general type also have equilibrium constants very close to 4.0. For a statistical calculation of the equilibrium constant for the above reaction, let it be given that: (i) the bond lengths in all three molecular species are essentially the same; (ii) in all three species the electronic ground state is nondegenerate; and (iii) the characteristic vibrational frequencies are as follows:

$$O^{16}O^{16}, \quad v = 4.7379 \times 10^{13} \text{ sec}^{-1}; \qquad O^{18}O^{18}, \quad v = 4.4669 \times 10^{13} \text{ sec}^{-1};$$
$$O^{16}O^{18}, \quad v = 4.6042 \times 10^{13} \text{ sec}^{-1}.$$

a) From the stated frequencies calculate $\Delta \epsilon_0^0$ for the reaction as written, and show that $\Delta \epsilon_0^0 / kT = 0.0029$.

b) Formulate the equilibrium constant for the reaction, using equation (30) supplemented by the definition expressed in equation (32). Perform all the simple cancellations that are possible. Note that, with all bond lengths equal, the ratio of I's can be reduced to a ratio of μ's.

c) Give a completely numerical expression for the equilibrium constant. *Without going through the many multiplications and divisions* indicated in your expression, you can easily pinpoint the origin of an equilibrium constant closely approximating 4.0. What is that origin?

31. Equation (33) reads:

$$\ln K_p = -\frac{\Delta E_0^0}{RT} + \frac{1}{R}\left[R \ln \frac{\prod (z_p^0/N)^p}{\prod (z_r^0/N)^r} \right].$$

Comparing this result with the general thermodynamic equation on p. 118 we concluded that, when ΔE_0^0 is a good *approximation* for ΔH_T^0, the expression in brackets above is an equally good *approximation* for ΔS_T^0. To gain some insight into the closeness of the approximation, consider an equilibrium mixture of ideal-gas units that have *only* translational degrees of freedom.

a) Making use of the Kirchhoff-law cycle sketched in Fig. 27, and of the basic definition of H in terms of E, show that:

$$\Delta H_T^0 = \Delta E_0^0 + \tfrac{5}{2}RT(\textstyle\sum p - \sum r),$$

where $\sum p$ and $\sum r$ respectively represent the total numbers of moles of products and reactants indicated by the stoichiometry of the reaction as written.

b) Substituting in (33) the result obtained in (a), show that rearrangement yields the equation:

$$\ln K_p = -\frac{\Delta H_T^0}{RT} + \frac{1}{R}\left[\textstyle\sum pR(\tfrac{5}{2} + \ln z_p^0/N) - \sum rR(\tfrac{5}{2} + \ln z_r^0/N) \right].$$

c) Comparing an equation on p. 71 with the result obtained in (b), show that the last expression in brackets above is accurately equal to ΔS_T^0. Then comment on the degree to which this expression will be well approximated by the expression in brackets in equation (33).

d) Show how the line of argument noted above can be extended to an equilibrium mixture of diatomic ideal-gas units that have *both* translational *and* rotational degrees of freedom.

Index